PRAISE FOR

THE CEO'S DIGITAL SURVIVAL GUIDE

"I have worked with Nathan for over twenty years in all facets of information technology, including cybersecurity, database management, and networking. He is, in essence, the soup to nuts in the IT world. Nathan is thorough in his research, the layout of a plan, working with team members on execution of the plan, and finally, a well-thought-through implementation."

—Colt Vollmann

President, VSR Industries

"If you own or are responsible for a business, you must read *The CEO's Digital Survival Guide*! The world of technology gets harder to navigate every year, but the reliance on it increases. Nathan has created a clear and easy-to-understand guide that condenses what a CEO needs to know to make the best risk-based decisions for their business's IT systems and information security."

—Leia Kupris Shilobod

CEO, InTech Solutions; author, Cyber Warfare: Protecting Your Business from Total Annihilation

"Nathan brings technical savvy and real-world know-how to business. His practical knowledge and professionalism keep networks running while eliminating security threats. Nathan's application of technology and thorough advice provide a powerful combination that significantly accelerates business growth."

—Brad Mishlove

CEO, Catapult Groups

"I have worked with Nathan Whittacre and his company for over twenty years. He has helped to guide us through the complex worlds of computers, networks, and, most importantly, security. He has a special ability to learn our needs and help us implement them in a safe, secure, and cost-effective manner. I have found that he has a special ability to explain complex technology issues in ways we can understand. I have called him the everyman's genius and I stand by that."

—Jim Hazle

System administrator, DOCS Management Services

THE CEO'S DIGITAL SURVIVAL GUIDE

THE CEO'S DIGITAL SURVIVAL GUIDE

A PRACTICAL HANDBOOK TO NAVIGATING THE FUTURE

Published by Advantage Books, Charleston, South Carolina.
An imprint of Advantage Media.

Printed in the United States of America.

10 9 8 7 6 5 4 3 2 1

ISBN: 978-1-64225-630-7 (Hardcover)
ISBN: 978-1-64225-629-1 (eBook)

Library of Congress Control Number: 2023907672

Book design by Analisa Smith.

To my children Nátali, Noah, Kathryn,
Sophia, and August.

You make me want to be the
best father I can be.

THE CEO'S DIGITAL SURVIVAL GUIDE

CONTENTS

ACKNOWLEDGMENTS

This book could not have been written without the influence and knowledge given by many people in my life. It is said that it takes a village to raise a child, and I often feel like a child around so many friends, family, and mentors who have led me to create this book. I extend my gratitude to all these people for guiding me over the years, sharing knowledge, and guiding me along my journey.

Family and Friends

Thank you to my parents, Jarrard and Ruth Whittacre, who taught me from a young age to work hard, learn as much as I can, and be persistent in achieving my goals. You taught me about entrepreneurship, finance, and sales. You had the vision to help a seventeen-year-old kid start a computer business. Who would do that? Thank you for trusting in me. You also taught me how to love and be loved and to be a compassionate leader. Most importantly, you taught me to love the gospel of Jesus Christ and live by his teachings. Thank you for helping me become the best person I can be.

I am eternally grateful to my *opa*, August Hug. Although you are no longer here with us, you were and always will be my hero. I always loved spending time with you playing games, helping you in your garden, or learning from your teachings. You had a love for life, and everyone around you felt it.

I dedicated this book to my children Nátali, Noah, Kathryn, Sophia, and August. I want to thank you for each day that I have with you. I feel that time is never enough, slipping between my fingers faster than I could ever want. Thank you for inspiring me to always try harder, be more understanding, laugh, and play like a champion each day. You all are my inspiration!

This book would never have happened without the love and support of my best friend, confidant, and fiancée, Joyce Forier. You have been my sounding board through the book writing process. Your encouragement to dedicate my time to write each day while I was going through one of the hardest times of my life kept me going. Thank you for reading, commenting, and implementing many of my recommendations in the book. You inspire me every day to be a better man and to find joy in the challenges of this life. I look forward to being together forever and for the many adventures we will have together.

Thank you to Ruth for the years spent together, helping me through my degrees, the fledging business, and achieving my goals. You gave us the greatest gift, our children.

To my siblings Tricia, Cyndie, and Brett, thank you for being the best family I could ask for.

Thank you, Tricia, for being my second mom. I fondly remember you taking me to the symphony and dancing to Disney music in your college apartment and the long talks we had when I was a teenager and visited you each summer. You are a wonderful mom to your beautiful daughters. Thank you for inspiring me to find joy in music.

Thank you, Cyndie, for teaching me how to be successful in business and have balance in life. I will always remember talking to one of your employees about how they looked up to you—a mom, business owner, and triathlete. I love experiencing all those things with you and hope to be still riding our bikes together for many more years.

Thank you, Brett, for being my business partner for many years. We built two great companies, and I couldn't be more proud of them. I've always looked up to you ever since we were little kids. Thank you for teaching me how to fix things. I love working with you, whether it is wrenching away at an old Mazda truck or programming complex database systems. Thank you for being a great big brother.

Thank you to my brothers-in-law Kelly and Larry and sister-in-law, Michelle, for always showing my family and me love. Thank you to my wonderful nieces and nephews, who were my first children before I had my own. I love my extended family!

I am grateful to the many friends who have inspired me in both my business and personal life. There are so many who I want to thank personally for your help, guidance, and love.

To my friends who have heard me talk about business way too much and given me more advice than I can count. Thank you to all for helping me along the way and picking me up when I fell along the trail of life.

Mentors and Teachers

Beyond my family, my greatest mentor in life was Linn Mills. Thank you for showing me love at a difficult time in my life and putting me back on the right course. Linn, you were a pillar in the Las Vegas community, with thousands attending your funeral, but you treated each person like your only friend. I'm still shaking your hand all these years later, thanking you for your guidance and compassion.

Thank you to Don Curry, my high school teacher and friend. I never took one of your classes, but our time together in Global Lab, lunch in your classroom, and all the other activities that you led taught me to love learning and love exploring the world. My first experience with the Internet was working with you to communicate to students throughout the world about science projects in our communities. Thank you, Mr. Curry, for opening my eyes to a whole new world.

I remember my first writing assignment in Saralyn Lasley's class. I thought I knew how to write an essay, but when the paper came back with red all over it, I realized that I had much to learn. Thank you, Ms. Lasley, for teaching me to love to write! Your class was always fun and entertaining. You inspired me to read, study, and learn about the world.

Being a nerd in school wasn't easy, and being called Doogie Howser, MD, by my teacher John Ball sure didn't help, but I knew it was a compliment to me. Thank you, Mr. Ball, for believing in me and my friend Jason Heffernan, stretching beyond our young abilities.

Thank you to my many college professors who taught me about technology and science, especially Dr. Yoohwan Kim, my master's degree examination chair; Dr. Evangelos Yfantis; and Dr. Laxmi Gewali.

My business life changed when I met Brad Mishlove in 2008. I knew very little about running a business at the time. Thank you, Brad, for always challenging me to grow, teaching me all things business, and being a good friend over the many years. It has been a challenging road, but you've taught me to always live another day. Thank you for introducing me to so many business leaders over the years who have guided me on my business (and life) journey.

Thank you to the rest of my Catapult Groups family for the advice and collaboration over the years. Thank you, John Kotek, Dave Lester, Darrell Evans, Nolan Jones, Matthew Kammeyer, Carlos Banchik, Adriana Gonorazky, Liz Ortenburger, Cory Summerhays, Nick Miller, Kent Bell, Jason Godfrey, Kevin Fults, Royal Marty, Joe Pantozzi, Brad Barth, Richard Bruno, and the many others who have come and gone from the group over the years.

Thank you, Robin Robins, for introducing me to Advantage Media and presenting me with the idea of writing a book. You inspire me, and all the TMT community, to be better business owners, stewards of the technology community, and better people. Thank you for helping me get off my butt and do all the right things to make my business successful.

Ross Brouse, a member of the TMT community, provided the concept and framework of the digital self-assessment used throughout the book.

Thank you to my friends in the TMT community and my Robot Rabbits group. Thank you, Bruce McCully, Vince Fung, Simon Fontaine, Ikram Massabini, Seana Fippin, Evan Desjardins, Greg Hanna, Mat Zoglio, Leia Shilobod, Peter Verlezza, Ryan Markel, Debi Bush, Richard Crockett, and Allison Foelber.

My Team, Clients, and Other Contributors

I often say that my team at Stimulus Technologies is my family. We started Stimulus as a family business, and I always want it to feel that way. We often spend more time together working to save the world through technology than we do with our own families. I have so many people to thank over the years that this section would be very long.

Thank you to every Stimulus family member who has been with me through the years. I sincerely appreciate your time and dedication to me, the company, and our clients.

I do want to thank a few people individually for their efforts.

First and foremost, thank you, Brian Alwood, for your dedicated service to the company and our friendship over the many years working together. We've had so many rubber duck sessions over the years, solving many critical problems and building a great company. Thank you for all the blood, sweat, and tears. We've been through so much, but you have always been at my side in building a great company.

Thank you to my leadership and management team for helping me guide the company to where it is today. I always appreciate the advice, pushback, and discussions in leading the company and the entire team in the direction we're going. Thank you, Kaid Whipple, Taryn Copeman, Khrystal Hill, Steve Goertz, Elaine Schaben, Ryan Herzing, Sherry Lipp, and Liz Branson. I appreciate our friendship and great working relationship.

Thank you to all the other Stimulus team members who have come and gone over the many years in business. One thing that I love is to see an individual become successful in their career, even if it isn't at Stimulus. May each of your journeys be to great heights and beautiful vistas!

Without our great clients and customers, Stimulus would not exist. I've developed great friendships over the years with many of them, and I call them my friends. Thank you, Jim Hazle and the entire DOCS team, for trusting us over the years. You've been a great friend and a true confidant. Thank you, Colt Vollmann and all the people at VSR Industries, for allowing us to create systems and processes to help your business. Thank you, Lila Foti and the Peninsula Fleet team, for giving me the opportunity to create software to run your operations and the friendship over the many years. There are so many other clients that I would like to name, but it would fill the whole book. Thank you for being a part of the Stimulus family!

Finally, thank you to Advantage Media for working with me to write this book. I sincerely appreciate the help from my writing coach, Suzanna de Boer. Thank you for keeping me on track, giving me advice along the way, and making the book better than I could imagine. I've appreciated our weekly calls and your ability to understand my technical writing. Thank you for making my words perfect.

ABOUT THE AUTHOR

Nathan Whittacre has always been passionate about technology and entrepreneurship. Originally from Salt Lake City, Utah, he moved to Las Vegas, Nevada, in 1988. He spent his early years reading about software programming and experimenting with BASIC programming on his home Atari computer. He founded his first business at age eleven when he started and ran a candy store for two years from his home. He was an early pioneer in telecommunications, running a bulletin board system for several years until he was introduced to the Internet. While in high school, he worked in several technology jobs building computers, helping maintain his school's computer network, and providing technical support at an Internet service provider.

Nathan founded Stimulus Technologies in 1995 with his brother and father with a goal to provide advanced technology solutions for his clients. His passion for innovation, analyzing hard issues, creating solutions, and fixing problems has created a portfolio with a wide variety of information technology skills. He earned his bachelor's and master's degrees in computer science from the University of Nevada, Las Vegas. As a lifetime student, he enjoys reading a diverse array of books on entrepreneurship, company culture, psychology, religion, technology, and science fiction. When Nathan is not building his company and helping his clients, he enjoys spending time with his five children, camping, running marathons, racing in triathlons, volunteering at church and scouting, and flying his airplane.

INTRODUCTION

KEEPING YOUR PLACE IN THE FUTURE

I have been working with companies for over twenty-eight years, helping them to be successful with the technology they need to ensure the security of their business. This usually entails building the right infrastructure, purchasing the right software, and protecting it with the right security systems. With almost three decades of experience, I've found business owners usually fall into one of three categories in regard to technology: gets it and wants it, doesn't get it but wants it, and doesn't get it and doesn't want to hear about it. My experience has shown me that the owners and their management teams that fall into the "gets it and wants it" category are usually the most successful in their business because they are students in all aspects of their business. They want to implement the best systems and technology in their company to make it easy and successful.

On the other hand, I am constantly called in to help companies that fall into the last category of "doesn't get it and doesn't want to hear about it." What has bothered me for many years is that I don't understand why these owners don't even want to hear about it. I go into these businesses and find ten-plus-year-old equipment, passwords written on sticky notes on laptops, systems that don't work, and employees who are frustrated. Usually, I get a call about the time that something has crashed or a hacker stole all their data and they can't work.

After each call and after I've worked to put their company back together, I've often asked myself, "Why?" We are always hearing about yet another company that has suffered from a cyberattack. These companies are big and small. From public companies and organizations such as Target, Equifax, the federal government, and Colonial Pipeline to small companies that don't get quite as much attention. These nonstop news stories *should* be a wake-up call to these owners to do something inside their companies. I've done countless seminars, webinars, blog posts, TV shows, and personal interviews, all talking about simple things that everyone, and especially business owners, should do to protect themselves. I've even talked to people after they have had an attack who told me they heard my talk but didn't think it could happen to them—until it did.

So where have I, and my industry, gone wrong? There are plenty of other books out there on cybersecurity. You could read an endless number of blogs and listen to podcasts about technology. After the last call I received a few months before writing this book, it really occurred to me where we have gone wrong as an industry. *It is just too complicated.* We use too many acronyms that people don't understand. We use "geek speak" and everyone's eyes gloss over. I also realize that I'm like a personal trainer or dietician trying to help people lose weight and get healthy. Everyone knows that they shouldn't eat fast food, but it is so easy to just drive through and grab a burger. Or why they should work out or go to the gym—because most people would prefer to do a Netflix marathon than run one. It is much easier to use fear to scare people to make changes than teach people about why they should do something good for themselves. Those are the tactics that I've used in the past and my industry generally uses too.

My concept for this book is different. I hope to teach you some things about technology, use some stories that are relatable, and give you tools to be able to implement them inside your company. More importantly, as we walk through these stories together, I hope to inspire you to make changes because you understand why it is so important to do it. Taking action with technology isn't about the technology; it is about making your business better and, in turn, making your life and the lives of those who work for you better.

Taking action with technology isn't about the technology; it is about making your business better and, in turn, making your life and the lives of those who work for you better.

You may feel overwhelmed or in a different world as I walk you through this journey. That is perfectly natural. Your core competency is in the industry that your business is in. If you feel that you can make this journey on your own, I know you will do great! Over the years I've read many books and implemented many new tools and systems in my business by reading a book, and the process of discovery and change was a fantastic experience. On the other hand, there were times that I felt like I needed a guide to help me through the process. I found the best coach, consultant, or team to help me improve my business. Just like hiring a guide to take you on a new journey in an unfamiliar land, there are many professionals who can help you through this journey. My firm, Stimulus Technologies, helps business owners and executives find success in their companies through technology. Whether you do this on your own, use a local guide, or contact me, I know this material will help your company grow in the new economy, be protected from all the threats that are out there, and help you to more easily achieve your company's security needs and goals.

My hope is that you will be able to take this journey with me to review your company's security requirements and, most importantly, enjoy the trip along the way!

My Reasons for Writing This Book

With over twenty-eight years of working with business owners and company executives on their security needs, I've had many experiences that have led me to write this book. The dangers are everywhere. I have found that most of the business leaders I work with really understand their business. Whether it is a doctor, dentist, home builder, CPA, car dealer, or engineer, they have all been very smart people in their respective industries. Most also know that technology will help their business and that they must use technology to gain competitive advantage. Most are also afraid of technology to some degree because they don't understand it as well as their primary line of business, and many technologists tend to talk above, around, and through them. They are often reminded of the old *Saturday Night Live* computer guy who would come into the office when the owner or employees were having issues, make some snarky remark, and just tell them to move so he can fix it. Many are also somewhat embarrassed that they don't know more about the technology their company is using.

I have three reasons for writing this book. First, to give you, all the business owners out there, a clear guide to help you speak and make decisions about technology as if it was your second language. Second, to give you a glimpse of how technology can help your businesses grow and operate more efficiently. Finally, to make it a really interesting book to read and not just a technical manual that only computer nerds would read.

If leaders of a business are not able to clearly understand technology or speak about it with an IT professional, it puts them at a disadvantage. Most executives are experienced professionals, have spent many years in college getting degrees—looking at you, my doctor friends—and are very smart and capable people. When a young twentysomething computer kid comes into their office talking about IP addresses, IPS firewalls, virtualization, or anything similar, they feel reluctant to admit that they have no idea what the guy is talking about. So they nod and allow them to do whatever they want. They also have issues making informed decisions about hiring a competent internal IT person or an external IT firm. They tend to allow problems or poor work performance to go on for months or years because they just don't want to deal with the issue, and they think all IT people are the same. My goal with this book is to give these owners the power to make good decisions about the technology in their company and have a reference guide to know enough about the technology they are using to lead their company into the future.

Technology allows companies to do more work faster and with less labor. Close your eyes for a minute and think about an engineering firm in the 1950s. (Okay, now open your eyes so you can continue reading.) Not just any engineering firm but a company that is going to design something that has never been done before. Imagine a large room filled with desks in neat rows, each desk with an engineer wearing a white shirt, tie, and horn-rimmed glasses sitting behind it. On the desk are a large roll of blueprints, a slide rule calculator, and a bunch of pencils. The engineers in that room are working day in and day out on designing a rocket that is going to send astronauts to the moon. I think everyone has a clear idea in their heads what that room looks like. Imagine how much time was spent erasing lines, throwing away paper, doing manual calculations, and just moving documents around from one person to another. That happened about sixty-five years ago, but it doesn't seem like that distant of a memory. If technology has advanced from metal slide rule calculators and pencils to high-speed, globally connected supercomputers that fit into our pockets in sixty-five years, then consider where it is going in the next sixty-five years. If company leaders aren't implementing current technology in their businesses, they will be left behind their competition that is implementing it correctly. Business leaders who do technology right will be able to build their businesses with less required labor and help those people work more efficiently, and everyone inside the company will be happier and more successful.

This is a book about technology. I want to admit something right now: I'm a nerd. I like reading textbooks. I generally open manuals and read the instructions. I geek out on the latest and greatest technology, whether it is about computers, cars, planes, bikes, or just about anything. I also know that I am an odd duck. Most people aren't like me. I know that there are hundreds, if not thousands, of books about cybersecurity, using technology in your business, or whatever latest book is out there on improving your company in many other ways. I've read a lot of them. Most don't address the CEO directly; instead, they are focused on the technologist who will implement the tools and information—and most of them are just *boring*. My goal with this book is to wrap some people stories from my life, including some of the fun adventures that I've been on, into a guide that will help the reader relate to the material. My hope is that the stories I present will make it easy to understand the acronyms and terminology so that you will remember them for years to come.

Guide to Using This Book

Have you ever used a trail guide to simplify and map out your journey? If you have, you know how much easier and safer your journey will be. I've written this book to be like a *trail guide*, which is broken up into different sections: food, shelter, first aid, etc. I have separated the book with each section having subsections that allow you to have quick references to review. There are different levels of understanding of technology; this is not a definitive guide on how to implement all the suggestions that I make in the book. View this book as recommendations that you can take to the professionals around you to help put in place in your business. Because businesses rely so much on technology today, I believe that the business owner, CEO, and anyone involved in major business decisions should have a fundamental understanding of how technology works inside the business. It should be one of your core competencies. As you implement these recommendations, you will see big increases in efficiencies. In other parts of the book, especially when dealing with security, it is important and essential to have all the recommended systems in place.

Many small business owners do not think that cybercriminals are after them or what they have on their computers and network. This big mistake can lead to your business losing tens or even hundreds of thousands of dollars, lost reputation, loss of clients, and potential fees and fines from the government. I recommend not just utilizing this handbook but also utilizing experts to help you implement technologies in your business. This is only a guide, and like many guides, your mileage may vary. I don't know all the circumstances around your business or industry. I recommend that you consult with your local trail guides and experts to help you along your journey. Having a general knowledge of what needs to be implemented inside your company will help you understand those experts better and allow you to make the best decisions about their services and recommendations.

Each chapter of the book has several parts.

An *introduction story* brings the material that I'll present in the chapter to life. My goal is to bring some experiences inside or out of technology that relate to the topic of the chapter.

Each chapter includes a section of *definitions*. There are so many acronyms in technology. Your technology professionals will throw these acronyms around like candy on Halloween. My goal is to provide you with a quick reference to these terms that technologists use all the time.

There are several subsections in each chapter. These subsections will give you a more detailed understanding about the main topic.

Finally, each chapter contains another self-assessment on topics discussed in the chapter. Use these assessments to identify where you are at currently and pick a subsection to focus on after reading that chapter.

My goal is for you to be able to use this book as a reference manual that you can go review over and over again as you improve and implement technology in your company.

There are guideposts along the way.

Trails generally have markings that help you find your way. You will see some graphics in this book to be a reference for you. Look for these guideposts to lead you on your journey.

Follow the guide: This is a tricky part of the journey. Be sure to have an expert on the trail to help you through this part of your journey.

Side trail: This is a beautiful trail to enhance your journey. This book gives you a taste of this information. For additional resources, visit my website: www.nathanwhittacre.com/resources. There are also references to other books and resources that I recommend.

Danger: This is an important precaution you should be familiar with. Not heeding this warning could lead you and your company off onto a dangerous trail.

Campsite: Take some time in this section to work with your team, assess your current situation, set goals for the journey, and eat some s'mores.

Trail junction: Sometimes trails have diverging points that lead you along two different trails, equally enjoyable. You just must make a decision on which direction makes the most sense for your journey.

Assess your location: Take some time to make sure you are on the right track. You may think you are on the main trail but end up going down a side path that leads to a dead end.

Although this book has serious material in it, I hope that your journey is memorable and enjoyable. This journey is about making your company better through technology. Enjoy the journey, and look forward to the goal!

CHAPTER 1

A DIGITAL SELF-ASSESSMENT

I often say that I didn't learn how to run a business in school. The designers of the computer science curriculum at my university must have forgotten to include a business management, accounting, finance, human resources, or any type of "running a business" survival class in between my programming, engineering, math, and science classes. To be able to run and grow a business over the past twenty-eight years, I've relied on three sources of knowledge and wisdom: trial and error (much more on the error side), consultants and other professionals, and books. I've listened to hundreds of great speakers over my time in business and read countless books about all aspects of running a business. One thing that always has helped me improve my business is doing a self-assessment about the topic—usually failing the assessment—and then using the knowledge from the book or speaker to improve a few aspects of how I operate my business.

With my experience in mind, I am going to start by giving you an assessment. This is not a test. There is no pass-fail for this assessment. Well, technically, there is a failure option because if you either don't have most of these things in place in your business or don't understand what they are, then your employees, the hackers, and your competition will test your business technology, and they will make sure your business fails. No pressure now—this is to see where you are at in your business and for you to find out what you do know and don't know about the terms and technology that we are going to be discussing in this book.

This assessment has a score between 30 and 90. You will rank each question between 1 and 3.

1—You don't understand the question, or you have not implemented it in any way in your company.

2—You have a general understanding of the question and have attempted to implement it in your company, but you know it is not working 100 percent for you.

3—You understand the information in the question, and you know that it is fully implemented across your organization.

Be honest with yourself. This is just the beginning of your journey. By the end of the book, you will understand these concepts and know who needs to implement them inside your company. For now, just get a score for yourself. This is the starting point. The assessment is divided up into six sections: infrastructure, cybersecurity, compliance, backup and disaster recovery, business strategy, and cloud. Each subsection has a possible score of 15.

If you don't want to write in the book and make this a permanent record, you can download this as a PDF from my website: https://nathanwhittacre.com/resources. Feel free to download it as many times as you want and also have your team take it. It is always interesting to see where your assessment compares against others on your management team.

INFRASTRUCTURE	SCORE
We maintain an inventory of all workstations, servers, and network equipment, and we have implemented a sustainable hardware refresh cycle.	
We utilize an incident ticketing system, and we provide our management team with regular response and resolution time reports and whether the results of those reports are meeting the organization's expectations.	
Our wireless network prevents guests from accessing our internal network, and employees have unique usernames and passwords assigned for wireless access.	
Our office locations utilize redundant Internet service provider connections, and our firewall or router automatically swaps connections in the event of an outage.	
Our servers and network equipment are protected with uninterruptible power supply units that are replaced every three years, are in a physically secure location, and have separate HVAC systems.	
Infrastructure Score **Total Possible 15**	

NOTES

Cybersecurity	SCORE
I am confident that we have the proper cybersecurity software deployed to protect personal and corporate data from attacks such as phishing and ransomware, including any remote devices that are used to access critical infrastructure and data.	
We engage with all organization employees and properly train them to identify ransomware, phishing, and social engineering attacks coming from emails, text messages, and websites.	
All organization IT systems and devices that contain personally identifiable information (PII) or sensitive company information are encrypted to protect against loss or theft.	
We use single sign-on and two-factor authentication across all critical lines of business applications, such as Office 365, our ERP system, and remote access.	
The level of cybersecurity insurance carried by our business is adequate to protect our organization and our clients from financial loss.	
Cybersecurity Score **Total Possible 15**	

NOTES

Compliance	SCORE
We apply regular server and workstation security patches and updates across our technology infrastructure.	
We have a properly segmented corporate network (meaning workstations, servers, phones, and guests are kept in separate logical networks).	
We perform a regular network vulnerability scan and have archived all historical scan data for reporting and compliance purposes.	
We have a written information security policy that has been agreed to by all employees.	
We are meeting all state and federal compliance requirements, such as HIPAA, PCI DSS, FINRA, and all federal and state PII rules, and we are confident we would pass an audit.	
Compliance Score **Total Possible 15**	

NOTES

Backup and Disaster Recovery	SCORE
We proactively monitor our server and cloud infrastructure for failures and performance issues so that business affecting problems can be prevented.	
We regularly review our backup strategy, and we adhere to a documented process for backup frequency, retention, and location.	
We perform regular backup recovery testing, and we have a clear time objective for restoring critical systems and data.	
Our management team understands and has agreed to the recovery time objective (RTO) and recovery point objective (RPO) for backup and disaster recovery, and both are monitored and tested regularly.	
We have a well-defined disaster response team with clearly defined roles, responsibilities, and communication protocols.	
Backup and Disaster Score **Total Possible 15**	

NOTES

Business Strategy	SCORE
The organization's management team views technology as an investment, not a cost, and they agree to implement best practices when recommended by the IT team.	
We perform a regular technical alignment assessment to identify areas of our technology infrastructure that do not meet best practices.	
We meet regularly as a team to assess risk, discuss strategy, and perform IT budget planning for our organization.	
We have a clear process for making IT-related decisions in our organization, a project plan is agreed upon before implementation, and communication within our organization is clear and consistent.	
We consistently bring advances in technology to the attention of our management team, which increases employee productivity and gives us an edge over our competitors.	
Business Strategy Score **Total Possible 15**	

NOTES

Cloud	SCORE
We utilize a secure cloud-based email solution, such as Microsoft 365.	
Our cloud services are configured according to the service provider's recommended best practices.	
Our cloud-based email and file services are configured with data loss prevention policies and alerting to prevent data breaches.	
Users have the ability to work wherever they are and whenever they want to work. The connection to the company's infrastructure is secure. The infrastructure can grow and shrink as business needs change.	
We utilize a 24/7 security operations center (SOC) that monitors and alerts on our network, cloud services and critical data systems.	
Cloud Score **Total Possible 15**	
NOTES	
TOTAL **Total Possible 90**	

Aren't you glad that is done? Don't worry, it is nearly impossible for an organization to have a perfect score. If you have 75 or above, you are doing good, and this book will help you improve on the foundation of technology that you have built. If your score is between 50 and 75, you are well on your way to understanding what you need to do but most likely need to build the infrastructure inside your organization to implement these pieces of your technology. If you are below 50, I'd suggest focusing on a particular area that you think you can get some quick progress on and start working on it.

At the end of each section of the book, you'll have an opportunity to take another self-assessment on what you've learned from that part of the book. This is a guide and a workbook. My goal is to help you effectuate change inside your business so that you are ready for the future.

CHAPTER 2

BUSINESS PLANNING AND ANALYSIS

My brother, my dad, and I started Stimulus Technologies in 1995. The midnineties was the dawn of the Internet, but most people hadn't even heard of it or were not using it. Just to put it in perspective, Google was founded in 1998, Facebook in 2004, and YouTube in 2005. Many of the technologies that we use each day did not exist. Moreover, my dad is old school. He earned his business administration degree in the early 1970s and ran several successful businesses through the 1970s and 1980s. I looked at him as my expert to help me set up the business and operate it. The partnership between my father, who had the experience of running a business, and my brother and me, who understood technology, gave us the tenacity to overcome many trials that would come.

We created Stimulus Technologies in my brother's garage out of used parts and doors made into workbenches and used handwritten ledger paper to keep track of payroll and inventory. I remember sitting down with my dad the first time we wanted to cut ourselves a paycheck. We manually calculated the required payroll taxes and created a handwritten paycheck that I was very excited to deposit into my bank account. I couldn't imagine doing that same process today with the seventy-plus employees that I currently have or for the thousands of employees that many companies employ. Whether it's just one person or hundreds or thousands of employees, running a business is complex. There are wonderful tools available today to help you run your business. This first section is about using some basic tools to start your business and how to make it run well.

On the next page is a list of key definitions that you can refer back to as you read this chapter.

Key Definitions

- **Rightsizing:** Ensuring that an organization has correctly scaled its software, services, and hardware.
- **Pro forma:** A tool that businesses use to map out their financial future. It incorporates a budget or financial forecast and is used to help map out different scenarios that a business could encounter.
- **Hardware life cycle:** The life span of the equipment that your company buys. Each computer, server, monitor, etc. has an estimated time to function normally.
- **Software life cycle:** The life span of the software that the company uses, usually dictated by the company that wrote the software and how long they will continue to provide support, updates, and security patches for the software.
- **End of support:** When both the hardware and software reach their end of vendor support. After this date, the manufacturer will no longer keep the spare parts of the hardware, release software or firmware updates, and no longer provide support for it.
- **Mean time before failure (MTBF):** The estimated time frame before the hardware is expected to fail on mean average by the manufacturer.
- **Technology leadership:** The management team of the company takes control of their technology, advances it to help their organization be the best it can be, and not let other companies, including vendors and competitors, dictate how it will use technology.

Information Tools

It really amazes me how far technology has come in such a short amount of time. If you think back historically, the transformation that has happened in this century *alone* is astounding. In 1965 Gordon Moore, one of the founders of Intel Corporation, postulated that every two years, the number of transistors on a microchip will double, and the price of the chip will be cut in half. Since he stated that theory, it has held true, and it has been called Moore's law. This means for you and me that the cost of computers will be cheaper over time, and they will consistently get faster and more capable.

Because of this increase in computing power, our access to information has also greatly increased. If you lived before the year 2000, think of how you planned a cross-country trip. What information did you use to get to your destination? How did you find out what types of places to visit along the way? How did you make reservations to sleep at a hotel or campground? How did you map out the roads you wanted to travel on? More than likely, you consulted friends or family who traveled that route before, bought a map or travel guide, and had to call for reservations on a landline phone. You probably carried the map(s) with you in the car and had to have a navigator help you find your way. You may have had to stop at a gas station and ask for directions if you got lost. It required planning, thought, and proper execution to succeed.

Think of the same trip today. You would probably use online tools, such as Google, Yelp, Tripadvisor, Travelocity, or any number of other sites, to research your destination, find the best routes, and book the places to stay. You would probably load up the maps on your mobile phone, and it will tell you exactly how long it will take to get to your next destination. You can find reviews about places to visit and make choices based on the number of stars that people rate the place. You could even post your plans on social media to see if anyone else you know has visited there and give you their ideas about the trip. The information available at our fingertips for our personal lives is amazing.

Similar to planning your trip, starting and running a small business today is, in some ways, much easier than it used to be. When I started Stimulus Technologies in 1995, even in the technology industry, much of my research, ordering parts, and other communication happened over the phone or through the mail. Many organizations hadn't adopted email or the Internet yet. To get information from a manufacturer about a computer part, we would have to use a modem to dial into their system and download the files. Today we just visit their website, look for what we need, and download the information.

A micro business today, with just one or two employees, has access to and can run a business with similar tools of a large company. These tools allow a small business to be nimble in the marketplace, move faster than a large company, and grow quickly. They also have access to the same information just like the Fortune 500 company. Armed with a cell phone, tablet, or laptop, a small business can provide services to their next-door neighbor or someone many thousands of miles away effectively. The tools that are available for that type of service and collaboration are incredible. As you continue with this guide, you will learn how to effectively leverage those tools to grow your small business and be as effective as any large corporation. Careful use of the tools and technology available will allow you to easily scale your business, be as productive, and eventually outmaneuver large businesses.

In the past it was thought that the advantage of being a small business is that it could be nimble and change quickly. Also, it was believed that the advantage of a large business is that it had the tools and capabilities to produce products and services much more effectively and efficiently through processes and systems. That was true in the past, but today a small business can be as effective as a large business. Using processes and systems along with small business tools, partners, outsourcing, and crowdsourcing to stay nimble, a small business can often be as powerful in the marketplace as a much larger company. It is all about mindset and using the tools effectively to run your business.

I'll share a few examples to illustrate this point.

Historically, if a construction company wanted to allow collaboration between teams for file sharing and project planning, they would need to set up several servers, VPNs, data replication across sites, and the security to protect the data. Although these systems were very capable, they were very rigid and expensive. Today tools such as Microsoft Teams, OneDrive, Office 365, Google Apps, and other tools allow teams to collaborate and work on data effectively and seamlessly. Multiple people can be editing the same document at once, sharing information back and forth easily. Additionally, these systems are provided as a cloud service, which means you can add or remove users to the system at will and don't have to pay all the up-front cost for the infrastructure.

Another example is accounting and financial systems. I mentioned that when I started my company, we processed the first payroll by hand, wrote checks, and mailed in the quarterly tax returns. I remember walking into the bank several times with my father and brother to set up our bank account and line of credit. All of that was very time-consuming and laborious. Today all that could be done in your pajamas at home. There are many great financial and accounting programs on the market. These have the same power for accounting and financial services that a Fortune 500 company would use but are as effective for small business. The advantage of these tools is that they scale easily as you grow, and you can pay for them monthly rather than having to go out and buy large packages.

I will discuss these tools in depth throughout the next chapters. The marketplace is always changing; therefore, use this book as a guide to the types of systems that are available, and find ones that work for you today.

Growth

Growth can be good and growth can be bad. Bigger can be good and bigger can be bad. "Grow or die" is a belief that has no basis in scientific research or in business reality. When not approached carefully, growth can destroy value as it outstrips a company's managerial capacity, processes, quality, and financial controls, or substantially dilutes customer value propositions.

—EDWARD D. HESS

Many business owners and executives believe that the company must grow, or it must die. It is true that without growth, a company is sliding backward in the marketplace. Because of inflation, if a company keeps prices the same and has the same revenue year over year, it is sliding back in sales. I believe that is where this belief comes from.

Taking a little side discussion away from technology, in many years running my business, I've been focused on top-line growth. I, too, believed that additional sales would bring greater prosperity. The problem with focusing solely on sales is that it can strip away from profitability, which is the actual core engine of the business. There's an old business fable that I always like to think about. One day an operations manager and a sales manager were sitting in a meeting. The operations manager complained, "We are losing money on every one of these parts that we produce. We need to raise prices." To which the sales manager responded, "Don't worry, we'll make it up in volume!" And that is the problem with growth—it can blind a business by the ego of the owners or executives by raising the top-line number and destroying the company in the process.

> **Implementing the right technology can increase your employees' productivity dramatically, allowing you to do more work for less labor cost. Your competitors are already doing it, so why aren't you?**

One of the best ways to grow is by implementing technology. Much of the world's wealth has come because of technological advancements. Through implementing new systems, processes, and technology and other innovations, we've been able to produce more goods and services with less labor and materials. In other words, if your company isn't implementing the latest technologies, you are losing out on profits that your competitors are probably capitalizing on. Rather than thinking about buying computers and software as a business expense, it is time you start thinking of technology as a way to grow your business and make more profits.

One quick example of this is the idea of dual or multiple monitors on a desk. It has been shown through studies that adding on additional monitors for users can increase productivity by up to 42 percent.[1] I find some companies unwilling to spend the extra $120 for a user to have a second monitor, even though that could be paid for in less than one day's productivity gains. The same thing goes for slow computers, outdated software, system outages, and other network problems. Implementing the right technology can increase your employees' productivity dramatically, allowing you to do more work for less labor cost. Your competitors are already doing it, so why aren't you?

Budgets and Pro Formas

It is much easier to look back and explain exactly what happened than it is to look forward and predict the future with certainty. Pulling historical financial reports for a business is simple and easy. Most accounting software can do it in a few clicks. If you have recorded the transactions that have occurred in your business correctly, the statements are accurate. Predicting the future is much more difficult. In the first few years of business, I found that I was wildly optimistic about the performance of the business. Just because I started the company, put an ad in the yellow pages (yes, that used to be a thing), and created a web page, business would flock to me. That proved to be very untrue.

In developing pro formas or budgets for the business, my dad gave me some great advice: cut your predicted sales expectations by at least half and double your costs. That will be closer to the truth than anything else. For a young business owner with great optimism,

1 Jon Peddie, *Jon Peddie Research: Multiple Displays can Increase Productivity by 42%*, October 16, 2017, https://www.jonpeddie.com/press-releases/jon-peddie-research-multiple-displays-can-increase-productivity-by-42/ (accessed October 12, 2022).

that was tough truth serum. Fortunately, I have learned that over the years it is very true and have put it in practice as I continue to grow the business.

Each year, and sometimes more often if the business is volatile or changing, I create at least a three-year pro forma. This is a bit more complex than a budget because it is designed to forecast sales, expenses, *cash flow*, and future balance sheet statements on all the important accounts. One of the most difficult parts of running a business is managing cash. Even a profitable business can run out of cash. Without cash, the business will stop operating.

For my pro formas, I continue to use Microsoft Excel, a great business tool. As I mentioned, when the business is volatile, this should be updated often, even monthly, to adjust future projections and allow you to make better business decisions faster and see the results of those potential decisions. When the business is more stable, it can be updated quarterly or even yearly. You should always look at your actual performance against the predicted performance to make future adjustments. The longer you are in business, the more accurate the forecast should be. There are a few times it could be dramatically off: during uncertain economic times (such as a recession), deployment of a new product line, or an unexpected loss of a large client. When those things happen, you should revert to updating the pro forma often to see what will happen if you make different decisions, such as hiring new people, buying equipment, moving offices, etc.

One of the things that I find very useful with a pro forma is running different scenarios. For example, let's say you are planning on hiring a new salesperson. You have an idea how much you want to pay the new superstar and the expectations for sales. You may want to make a few copies of the pro forma and run some different scenarios. Does it make sense to enhance the person's base salary or the commission structure? How many sales does the new employee need to make to break even? What effect does it have on your operations? Will you need to hire additional staff to handle the new workload? Maybe the answer is that growth that occurs too quickly may cause a cash shortfall, unless you have plenty of reserves to be able to handle that additional growth. These are all questions that you can answer with a good pro forma.

	2022	2023	2024
Ordinary Income/Expense			
Income			
Nevada Sales	$2,125,000	$2,550,000	$3,060,000
California Sales	$4,525,000	$5,430,000	$6,516,000
Total Income	$6,650,000	$7,980,000	$9,576,000
Cost of Goods Sold			
Nevada COGS	$1,062,500	$1,275,000	$1,530,000
California COGS	$2,262,500	$2,715,000	$3,258,000
Total COGS	$3,325,000	$3,990,000	$4,788,000
SubTotal Nevada Gross Profit	$1,062,500	$1,275,000	$1,530,000
SubTotal California Gross Profit	$2,262,500	$2,715,000	$3,258,000
Gross Profit	$3,325,000	$3,990,000	$4,788,000
Nevada Expense			
Advertising	$63,750	$76,500	$91,800
Automobile Expense	$42,500	$51,000	$61,200
Bank Service Charges	$53,125	$63,750	$76,500
Employee Benefits	$106,250	$127,500	$153,000
Insurance	$40,000	$45,000	$47,500
Licenses and Permits	$1,500	$1,750	$2,000
Office Supplies	$35,000	$37,500	$40,000
Payroll Expenses	$39,000	$42,250	$45,500
Rent	$65,000	$66,500	$67,500
Salaries and Wages	$300,000	$325,000	$350,000

Taxes	$12,000	$13,000	$14,000
Telephone	$7,500	$8,500	$9,500
Training	$5,000	$6,000	$7,000
Travel and Entertainment	$6,500	$7,500	$8,500
Utilities	$6,000	$7,000	$8,000
SubTotal Nevada Expenses	$783,125	$878,750	$982,000
California Expenses			
Advertising	$135,750	$162,900	$195,480
Automobile Expense	$90,500	$108,600	$130,320
Bank Service Charges	$113,125	$135,750	$162,900
Employee Benefits	$226,250	$271,500	$325,800
Insurance	$60,000	$65,000	$70,000
Licenses and Permits	$2,500	$2,750	$3,000
Office Supplies	$35,000	$37,500	$40,000
Payroll Expenses	$97,500	$110,500	$123,500
Rent	$85,000	$95,000	$100,000
Salaries and Wages	$750,000	$850,000	$950,000
Taxes	$17,500	$20,000	$22,500
Telephone	$10,000	$12,500	$15,000
Training	$7,500	$8,500	$9,500
Travel and Entertainment	$6,500	$7,500	$8,500
Utilities	$7,500	$8,500	$9,500
SubTotal California Expenses	$1,644,625	$1,896,500	$2,166,000
Total Expense	$2,427,750	$2,775,250	$3,148,000
SubTotal EBITDA			
Nevada	$279,375	$396,250	$548,000
California	$617,875	$818,500	$1,092,000
EBITDA	$897,250	$1,214,750	$1,640,000
Other Income			
Other Nevada Income	$7,500	$8,500	$4,500
Other California Income	$2,750	$1,500	$5,000
Total Other Income	$10,250	$10,000	$9,500
Other Expenses			
Other Nevada Expenses	$45,000	$35,000	$25,000

Other California Expenses	$75,000	$65,000	$55,000
Total Other Expenses	$120,000	$100,000	$80,000
SubTotal Net Profit			
Nevada	$241,875	$369,750	$527,500
California	$545,625	$755,000	$1,042,000
Net Profit	$787,500	$1,124,750	$1,569,500
Cash Flow			
Principal Payments			
Nevada Payments	$24,000	$26,000	$28,000
California Payments	$36,000	$39,000	$41,000
Total Principal Payments	$60,000	$65,000	$69,000
Other Cash Items			
Nevada Other Cash Items	$75,000	$125,000	$150,000
California Other Cash Items	$125,000	$200,000	$250,000
SubTotal Other Items	$200,000	$325,000	$400,000
Subtotal Nevada Cash Flow	$142,875	$218,750	$349,500
Subtotal California Cash Flow	$384,625	$516,000	$751,000
Net Cash Flow	$527,500	$734,750	$1,100,500
Nevada Cash Balance	$142,875	$361,625	$711,125
California Cash Balance	$384,625	$900,625	$1,651,625
Total Cash Balance	$527,500	$1,262,250	$2,362,750

Long Term Liabilities			
Nevada Long Term Liabilities	$250,000	$226,000	$200,000
California Long Term Liabilities	$450,000	$414,000	$375,000
Total Long Term Liabilities	$700,000	$640,000	$575,000

I would suggest that business owners or CEOs should be very involved in this process or be the one to run the pro formas. It gives them the sense of what is going on in the business, and as they look at the numbers in detail, it gives them a chance to understand where money is going in the business. It is also a check and balance against possible fraud in the organization. By looking at the numbers in detail, the business executive can get "into the weeds" for a few minutes to see if the numbers aren't as they expected or if money is being spent where it shouldn't. One of the major themes of this book is checks and balances. Make sure that no one person or group of people have control over your business. It is essential that you, as the business owner or executive, understand all the aspects of the business so that no one employee feels that they have an opportunity to do something they shouldn't. It also enables the owner or executive to make the best decisions possible for the business.

Controlling the Hardware and Software Life Cycle

From my experience most business owners want to use their hardware and software as long as possible before replacing it. They generally do not have a plan to replace or maintain the technology system that they own and only replace it when it breaks. The best analogy that I can compare this to is vehicles. Under this same premise, the company would buy a fleet of vehicles and only do maintenance when the check engine light comes on or a tire blows out while driving. They would only replace the vehicles when they got stuck on the side of the road, when they were towed back, and when the mechanic told them it would cost more to repair the vehicle than replace it. It is definitely a way to manage vehicles, but it will leave the people driving them frustrated with all the downtime along the side of the road as well as worried about their safety while driving.

Keeping up to date on software and hardware is the first key to maintaining a secure technology environment inside the business.

Keeping up to date on software and hardware is the first key to maintaining a secure technology environment inside the business. There are several components required to start to get control of this life cycle.

- Maintain an inventory of all hardware, including make, model, serial number, purchase date, warranty status, and who the device is assigned to.
- Maintain an inventory of all your line-of-business applications, including the vendor, installed version, support agreements, and support expiration.
- Set a hardware replacement life cycle for each of your types of devices: tablets, laptops, desktops, and servers.
- Replace devices before their operating systems are "end of life" by the software company.
- Maintain current support agreements for all your line-of-business applications.

MAINTAINING A HARDWARE INVENTORY

Throughout this book I am going to tell you that there are many ways you can accomplish a recommendation. These methods can range from very basic to advanced. What is important is that you do something that can be maintained easily. We'll get into some detail about types of systems you can use to document the policies and procedures to maintain these lists in chapter 12.

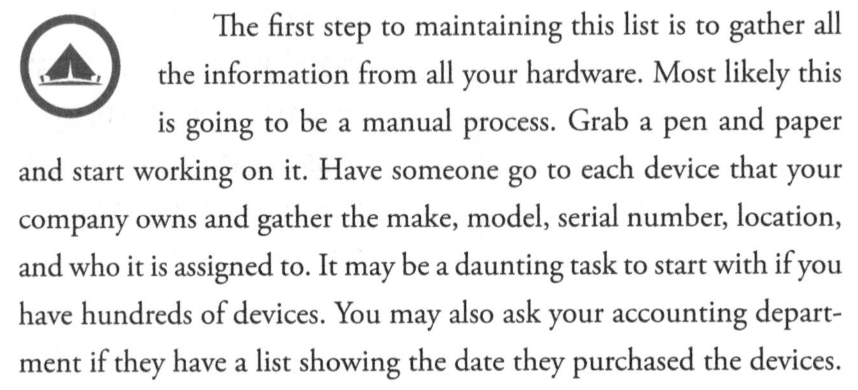

The first step to maintaining this list is to gather all the information from all your hardware. Most likely this is going to be a manual process. Grab a pen and paper and start working on it. Have someone go to each device that your company owns and gather the make, model, serial number, location, and who it is assigned to. It may be a daunting task to start with if you have hundreds of devices. You may also ask your accounting department if they have a list showing the date they purchased the devices.

To get the warranty status and purchase date, most manufacturers have a website you can go to find the information, such as the following:

Dell: https://www.dell.com/support/home/en-us?app=warranty

HP: https://support.hp.com/us-en/checkwarranty

Lenovo: https://pcsupport.lenovo.com/us/en/warrantylookup#/

These links change over time. You can visit the manufacturer's main website for additional information. Once you have the list, you can decide to keep it on a piece of paper, so long as you are willing to maintain it on there. Just don't let your dog eat it. Minimally, I would suggest putting it in a spreadsheet. You can download the spreadsheet from my site: https://nathanwhittacre.com/resources. The most important part of this is to keep it up to date.

The more advanced version of this is to use software to maintain a list. For my clients we use a hardware life cycle system called ScalePad, which integrates with our management tools and automatically gathers and maintains this information. It even sends us and our clients a monthly report of their inventory, warranty status, potentially expiring warranties, age of the hardware, and hardware that is slow. It automatically updates the information throughout the month and collects new assets as they are added into the network. It makes tracking and maintaining the hardware inventory much easier. Anything you can do to automate the maintenance of this process, the easier it will be, and more likely it will stay current.

MAINTAINING A SOFTWARE INVENTORY

Along the same lines of maintaining a hardware inventory, it is important to keep track of your software licensing. This process used to be more difficult when software was generally purchased with a device and had a license key written on a box. In the past, companies generally kept their software boxes in a closet with the name of the machine written on it. It became difficult to keep close track of which computers actually had the software installed, and it was too easy to install the same software on multiple computers.

I want to share with you a quick and scary story from a few years ago. I know of a firm that kept track of their software inventory this way. They had several lines of business programs that were quite expensive, ranging from $500 to $5,000 per computer. The management team decided to buy software for the computers that they felt had an essential business need. The employees got frustrated with not having all the programs they wanted, went to the software inventory closet, and installed the software they wanted on their computers, with management unaware of what was happening. A few years passed, and this became a regular practice among about thirty employees. At one point an employee left the company and decided to report the firm to the Business Software Alliance (www.bsa.org), which pays whistleblowers for reporting illegal software use. The BSA reports that software companies are losing $46 billion a year (Compliance Solutions 2021)[2] in lost revenue, and their goal is to help those companies collect the lost revenue. The company went under a voluntary audit from the BSA, and they found that the violations and damages totaled over $250,000. The company had to pay that fine and also become current on all software going forward to avoid a lawsuit from the software vendors.

As the one ultimately liable for software licensing inside your organization, it is important that you clearly define what software you will use *and* who has access to it. In the "Internal Cybersecurity" chapter, we'll discuss how to prevent employees from installing software on the computers themselves and lead you to potential violations like the firm that had to pay $250,000.

Software subscriptions make this inventory process much easier. I understand that many business owners do not like the idea of paying a monthly fee to use software. However, the idea that you can buy software once and use it forever is not realistic. Software has an expiration on it, just like hardware. Most software companies price their software subscriptions about the same as if you were to purchase the software and use it under a normal life cycle. If you have a legitimate desire to maintain software to current standards and revisions, paying for a software subscription is not a bad thing. Many business owners push back on the concept; however, I encourage you to embrace it. It keeps you current on security patches, revisions, and support and makes license management easier.

However you choose to manage your licenses, you should keep track of subscription or license information and vendor information, including contact details, account numbers, and how to submit tickets or issues. It is also

2 Business Software Alliance, Compliance Solutions, accessed December 23, 2021, https://www.bsa.org/compliance-solutions.

important to make sure there are multiple contacts or administrative users from your end of the agreement so you don't lose access to their systems if the only person who has administrative rights leaves the company. More importantly, if someone leaves the company that is an administrator, you want to make sure you block their access as they leave so they can't get access to it after departure.

SETTING A HARDWARE LIFE CYCLE

Most companies have many different devices in their environment, from tablets and desktops to servers and complex network infrastructure. All this hardware has an estimated useful lifetime. Although it may last beyond that lifetime, it usually has more problems after a certain point.

Have you ever owned a car that just after the warranty expires, you start having major required repairs? Most car owners get very upset at the manufacturer saying that they set the warranty period so that the car will have issues just after it runs out. Why do you think the manufacturers set a certain period of warranty? Or when they sell you an extended warranty, it gets much more expensive when you increase the mileage or time span? The manufacturers know from experience how long major parts will last from a long history of tracking them. They sell the warranty knowing that statistically the vehicle will have little to no issues during that period, except for potentially manufacturing defects. After that period, things will start to fail.

In the computer world, there are components in devices that have the same type of life span. Historically, the most common hardware failures in computers are hard drives, fans, and power supplies. All these have certain amounts of mechanical or high-heat-producing parts and fail because of heat, wear, or dirt. Most computers today have more solid-state devices (meaning that there isn't a mechanical component to them), but even these have a fixed life span. For example, solid-state drives (SSDs) have a certain number of read and write cycles before they start to fail. Laptops, tablets, and other mobile devices have batteries that degrade over time. If you want to be really particular, you can look up the manufacturer's estimated mean time between failure (MTBF) of each of the components and figure out an estimate life span of each device. But as the popular Internet meme says, "Ain't nobody got time for that." Here is my recommendation for replacing your equipment:

TYPE OF EQUIPMENT	REPLACEMENT TIME FRAME
Laptops, tablets, and other mobile devices	three years
Desktops and workstations	three to five years
Routers and firewalls	five years
Servers	five to seven years
Peripherals (monitors, printers, etc.)	seven to ten years
Networking equipment	seven to ten years

I also encourage you to carry a warranty and hardware service plan on most of your important equipment for the anticipated hardware life cycle, especially servers. For mobile devices, desktops, or workstations, keeping spare devices on hand for quick replacement is essential for most businesses. The number of backup devices depends on the size of your company and the number of different types of devices you operate inside your company. For a rough number, keep a spare device for every twenty employees that you have. That way, if a computer fails, the employee can quickly get back up and running with little downtime. Remember, the manufacturer has told you that the equipment is going to die and approximately when. Any use you get out of the equipment past that date is a blessing but is also playing against the odds. Having a plan beforehand puts the odds in your favor.

BUDGETING FOR IT

I often ask this important question when interviewing new potential clients: "How do you budget for IT services?" The answer I generally receive is, "We don't." How would you answer that question? Do you know how much you plan on paying for technology in the next year? If you hire more employees, do you know how much it is going to cost you to purchase hardware, software, and add-on subscriptions for them? Technology should be part of your business's budget because it can account for a decent operating cost. More importantly, I've found that companies that take technology seriously are usually ahead of their competitors in service delivery, sales, and growth. It is an essential component to running your company.

One of the common budgeting scenarios is to wait for things to break or replace devices when you are forced to. That is like driving your car into the ground. It is an option and can work but often revolves around replacing systems all at once and sometimes unexpectedly. Over the years Microsoft has stopped supporting and patching their operating systems. Other software vendors follow suit and won't allow their software to run on expired Microsoft Windows operating systems. When that expiration date comes, it requires all computers to be upgraded or replaced, often at once. This can be a huge expense in one shot to the company.

I'm going to teach you a better way.

Once you have your plan for your hardware and software life cycle, know how much you plan on paying for software licensing and subscriptions, and have a detailed hardware and software inventory, you can effectively budget for IT inside your organization. It is so much better to know how much you are going to pay for services throughout the year rather than have it be a surprise. When you start this process, you may be surprised how old some of your equipment is. Create a spreadsheet, or have someone inside your company that is great with Excel create it, with all your equipment, purchase dates, and warranty expirations. Categorize the equipment by the categories that I detailed above for your hardware life cycle plan. Once you have that in place, see how much equipment you have that is currently outside your set hardware life cycle. If you do not have any, you're in great shape! If you do have some or most that are out of the life cycle, create a plan to replace them. Keep it within the budget of your organization, and try not to be disruptive to your employees. It may be best to limit your replacement to two machines per month for smaller companies (under twenty employees) and to five machines per month for companies with twenty-one to one hundred employees. Larger companies can look at department by department to make those choices of replacement. Keep this cycle going until all your machines are compliant, and then keep it updated in your yearly budgeting cycle. (You do this, right?)

For your software budgets, if you do not currently have support agreements on your software, contact the vendors to see how much it will cost to add back the support agreements and get current. Ask them if there is a subscription model so that you can stay current with them going forward. In the end the subscription model can save you money because it keeps your support agreement current, maintains the software at the latest version, and makes budgeting much easier because you can pay monthly or annually for the software rather than having to pay for it in one lump sum.

Business Planning and Analysis Checkup

Just like the initial checklist, give yourself a score of 1 to 3 in each area. Add them up to see how you did overall in this area. Focus on just one area to improve upon for now.

AREA OF CONCERN	SCORE
I meet with my key management at least annually to create and update the pro forma.	
I review the pro forma with my management team at least quarterly and adjust it based on business decisions that I am currently making.	
I keep an inventory list of all my hardware and software.	
I have a current list of all my software vendors, including support contact numbers, and a list of authorized users.	
I have a defined hardware and software life cycle, and all my devices are compliant with the life cycle.	
TOTAL **TOTAL POSSIBLE 15**	
NOTES	

Business Planning and Analysis Key Points

- Implementing the right technology inside your business can help it grow faster and be more competitive.
- You should be familiar with the technology your company uses and be the driver of innovation inside your organization.
- Hardware and software have expiration dates and should be updated and maintained regularly.
- Defining the right policies and procedures inside your company will lead to a proactive approach to running it rather than being reactive.
- The use of pro formas, system life cycles, and forecasting can help you plan for the future of your organization and analyze the results of decisions you are making today.

CHAPTER 3

INTERNAL CYBERSECURITY—PROTECTION FROM INSIDE THREATS

Planning for a long thru-hike requires significant foresight into all the different things that can happen along the way. Unfortunately, many of the possibilities of failure are internal, not external. My greatest enemy of having a great hike was myself. With the right mindset, though, most obstacles can be overcome. I thought, if it doesn't kill me, it should make me stronger.

The first big internal threat to a long hike is hunger. On the John Muir Trail, hikers are in the backcountry with no access to towns or places to get food for many days at a time. Some hikers fish in the alpine lakes, but the fish there are small and don't provide enough calories to sustain the calorie demands of a long hike. Additionally, all food is required to be carried in bear-proof canisters, which add weight and complexity. The largest canisters correctly packed with the right food can carry about ten days of food, so that makes planning a resupply of food important for the number of days it will take to hike it. The average hiker takes between twenty and thirty days to complete the hike, which means one to three resupplies along the way. The food needs to be not only calorie dense but also nutritious. If it wasn't for needing proper nutrition, a hiker would just carry a bag of olive oil and drink it for each meal. Lots of calories but not the right nutrition. And that would just be gross too! Also, proper food preparation is important. A hiker needs to carry the right type of stove that is very lightweight to be able to cook the food. Most turn to using freeze-dried meals that can be reconstituted with boiling water in about ten minutes.

Planning for potential injury is also difficult. Rangers will tell hikers that they have to self-rescue, if at all possible, in the event of an injury. Getting to hikers in the backcountry can take hours or days. Search and rescue isn't going to pick up a hiker because of a blister or sprain. It has to be a very serious injury. But there is a balance between carrying an entire EMT bag, which would weigh a ton, and too little, like just a bandage and an aspirin. Also, doing everything possible to avoid injury, such as having the right gear, being properly trained for the distance, and being aware of the dangers that are constant.

Fatigue is another big factor that faces a thru-hiker. Being away from home, walking all day, and not sleeping well at night wears on a hiker. When I hiked the John Muir Trail, I did it solo, which sometimes led to boredom or loneliness. Being aware of these possibilities and finding ways to overcome them is an internal battle. I found having an audio book for part of the hike, writing in my journal at night, and taking in the beauty of the surroundings to be good ways to overcome the pain that I felt from my legs or in my head from being homesick or lonely.

Being keenly aware of these difficulties beforehand helped me have a successful hike. No one checked my bags before going to make sure I had the right food. I didn't have a partner to push me when I was down. The success was on me, and I knew that going into the hike. Having a successful business is much like that. Your biggest enemy is from within, especially as you grow and add employees. Even well-meaning employees are a big threat to your company. In this chapter we'll discuss how to properly prepare against internal threats that could slow your growth or even blow up your business. You'll learn the required strategies to mitigate against the threats from within.

On the next page is a list of key definitions that you can refer back to as you read this chapter.

Key Definitions

- **Malware:** Software, such as viruses, or programs that run on a computer, often unknown to the computer's users to infect, monitor, corrupt, or otherwise cause the computer to act in a way the user does not intend. This can also involve collecting data from the computer to be sent back to a person or group that should not be allowed to access the system (i.e., hackers).
- **Antivirus:** Uses heuristics and databases to detect malware and viruses in the endpoint user.
- **Endpoint detection and response (EDR):** Monitors and detects suspicious behavior and activities at network endpoints.
- **Personally identifiable information (PII):** Information that can be linked to an individual either directly or indirectly.
- **Acceptable use policy (AUP):** Defines what your users can and cannot do on your computers and network.
- **Virtual local area network (VLAN):** Allows a network engineer to segment a network to protect critical infrastructure from other devices.
- **Active Directory:** A system that is built into Windows servers and computers that allows for central management of the devices.
- **Group Policy Object (GPO):** A definition inside the Active Directory that allows an administrator to set a policy for the network.

- **Local administrative rights:** When a user has permission to install, change, or remove software or access critical areas of the computer.
- **Zero trust:** The default security stance of the system or network is to not allow access to undefined programs, access, or communications.
- **Domain Name System (DNS):** A network tool that translates an Internet name (such as www.google.com) to an address where it can be found by the computer.
- **Filtering:** Creating a list of allowed sites or programs that can be accessed. This can be a blacklist—a list of sites that aren't allowed but everything else is—or a white list—a list of sites that are allowed, but everything else is blocked.

Layered Protection

If security were easy, a network administrator could install one device or one piece of software to provide all the protection a business needs. Unfortunately, there is no such thing as a magic network protection device that will stop all the hackers, prevent all viruses, stop your employees from stealing your data, and keep you safe from all the other threats, both inside and outside your network.

When planning, designing, and implementing network security, a network or system administrator has to think about layered protection. Each piece of software, security device, or training program has a specific target or type of attack that it will prevent. These different systems should all work in harmony among themselves to provide the protection that you need. Additionally, they need to be rightsized for your organization, even though *all* organizations need a minimum level of protection to operate today.

IDENTIFYING YOUR ASSETS

The first step in designing a system to protect your company is to decide what assets you need to protect. You can think of it this way: the protection required for someone storing millions of dollars' worth of fine art is very different from an individual protecting a few pieces of expensive jewelry in their home. Both the size and type of protection mechanisms are different. In the center of your layers of protection are your assets. In the business world, this may include the following:

- Banking and financial information
- Passwords
- Email and other online communication systems

- Voice over Internet Protocol (VoIP) and other phone systems
- Customer lists
- Stored customer information, such as credit card numbers
- Proprietary processes and methods of doing business
- Line-of-business applications and their data
- Regulated PII, such as healthcare information or financial data

Even small businesses have a significant number of assets that they may need to protect. Even a one-person business generally has several of these assets that they need to keep safe to prevent financial or business function disasters. Get with your team (even if your team is just you) and brainstorm about all the assets that you need to protect. In this process, ask yourself the following questions:

- What systems or information, if lost, would stop me from doing my day-to-day business?
- What information, if lost or stolen, would cause my business harm with my customers and the market?
- What information, if lost or stolen, would cause a problem with a governmental regulation?
- What would my competitor want to know about my business to cause me harm or put me out of business?

As you think about these questions, you can also put them into different buckets.

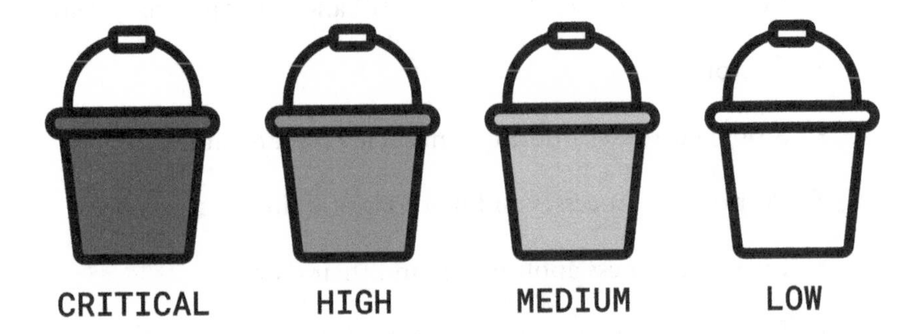

- **Critical:** Assets that, if lost or inoperative, would put you out of business.
- **High:** Assets that, if lost or inoperative, would cause harm to your business but you could recover from.
- **Medium:** Assets that, if lost or inoperative, would be detrimental to the organization, but you could continue operations with some changes.
- **Low:** Assets that are nice to have to make the business run smoothly but are not critical to the day-to-day function of the company.

When making decisions about security, you need to make sure your high and critical assets are protected before anything else. The buckets may not always seem obvious. For example, you may think that email is a low or medium asset. It would be disruptive to lose, but your business could continue to operate around it. Unfortunately, many attacks by hackers use email communications to infiltrate the company. A stolen email account can be used by the hacker to pretend to be an executive inside the organization and perpetrate wire fraud and significant financial loss for the organization. We'll get into that more in the next chapter. For now, make your list and try to think about which bucket each asset belongs in.

LAYERS OF SECURITY

Today the common model of security involves seven layers, starting from the inside out.

1. Assets
2. Data
3. Application
4. Endpoint
5. Network
6. Perimeter
7. Human

Each of these layers requires protection against inside and outside threats. For example, in the human layer, social engineering attacks by hackers would be considered an outside threat, which we'll cover in the next chapter. Whereas a disgruntled employee inside your organization can also cause harm by stealing confidential data and selling that information to competitors. In this chapter we'll look at ways to protect yourself from those types of insider threats.

Partitioning Your Systems

When I go into a company for the first time and analyze their systems and network, one of the critical things I look for is how their file systems are set up and who has access to what information. It is always interesting to me to analyze the information with the business owner because they are often surprised to learn which employees have access to different systems. One afternoon I was visiting a potential client who assured me that their file servers were protected and that only the owners had access to the financial data. After running my security scans on the network, I found that the way the previous administrator had set up the system allowed *all* employees to access the file share containing the confidential data. I shared that with the owner, and he didn't believe me. We decided to go to a random employee's computer and I was able to browse to the file server and gain access to the confidential shared drive. It would have required that employee to have the knowledge to go to the file server and look for that data, but it was there and accessible. Worse still, if my scan found that data, a crypto-locker virus on the network would have also been able to encrypt all that data and cause significant harm to the organization.

I also often find companies storing all their data in a single shared drive named something like "Public Files," and everyone has access to that. Upon a quick scan of the "Everyone has access to this data" folder, I find that individual employees create personal folders inside that drive where they are storing all their data, including often confidential financial information and critical password lists right next to their photos of their personal family vacation.

PARTITIONING YOUR FILE SYSTEMS

The easy first step to protecting your data is to create a new file structure for your data. Even in small organizations, not all employees need or should have access to all information. I would generally recommend some type of separation along these lines:

- General company information (i.e., public drive)
- Shared customer information
- Marketing
- Sales

- Administration
- Accounting
- Human resources
- Executive

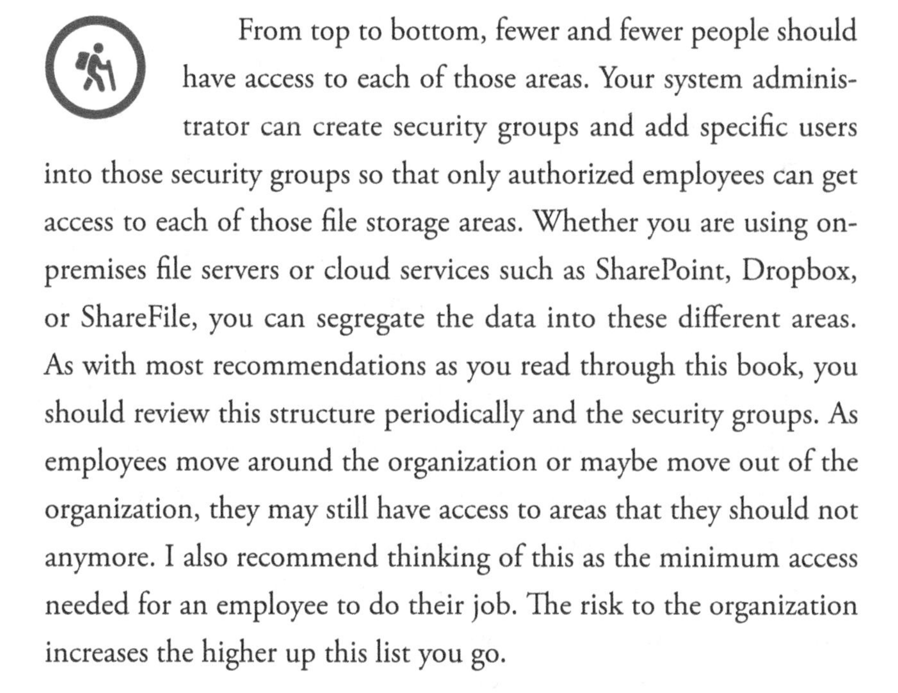

From top to bottom, fewer and fewer people should have access to each of those areas. Your system administrator can create security groups and add specific users into those security groups so that only authorized employees can get access to each of those file storage areas. Whether you are using on-premises file servers or cloud services such as SharePoint, Dropbox, or ShareFile, you can segregate the data into these different areas. As with most recommendations as you read through this book, you should review this structure periodically and the security groups. As employees move around the organization or maybe move out of the organization, they may still have access to areas that they should not anymore. I also recommend thinking of this as the minimum access needed for an employee to do their job. The risk to the organization increases the higher up this list you go.

PARTITIONING YOUR NETWORK

For the last few years, I have been going to a networking meeting once a month held in the offices of a local business. Ironically, I did a network analysis and proposal to this company quite a few years ago, and they declined my services. While at their offices, I decided to jump onto their guest wireless network. The passcode for the network was displayed prominently on the conference room wall and was their business phone number. Upon logging into their network, I had to accept the terms and conditions of their free Wi-Fi service (which is actually a good step), and it asked me for my name and email address. Unfortunately, that was when things turned south. I was surprised to note that I was connected to their general office network, which included access to their workstations, file servers, and printers. I wasn't attempting to access the network, but generally, when a Windows computer connects to a new network, it attempts to find its way around. If I were a bad hacker, I could have potentially brute-forced my way into their network systems and gained access to confidential company data.

Similar to partitioning your data, your internal network should be partitioned so that only those who need access to certain areas of the network should get that access.

Similar to partitioning your data, your internal network should be partitioned so that only those who need access to certain areas of the network should get that access. For most small organizations, that may be only a few separate networks.

- Office computers and servers
- Credit card processing
- VoIP phone system
- Surveillance/camera systems
- Guest network

This list could grow—for example, by department for your office computers—or be smaller—if you don't have VoIP phones or cameras. The concept of partitioning your network is that you can create layers of protection against critical systems so that if one system is compromised by an employee (internal threat) or hacker (external threat), the other areas of the network are protected. Additionally, it makes it harder for an employee who may just be snooping around to see what they can access to find any information that might be "interesting" to them.

An additional benefit to having partitioning, especially around regulated data such as credit card processing, is that it makes protecting that data simpler. You can put your "big guns" of protection around a smaller part of the network rather than having to implement the tightest security systems around every part of your network. Payment card industry (PCI) compliance requirements are often very stringent, especially if you are processing large dollar amounts. As a result the network requirements around those computers and systems that run credit cards could make it difficult for a regular employee to do their work with the same level of protection in place.

Acceptable Use Policy

Wouldn't it just be easier if technology could fix all your problems with technology? It is common for me to discuss the issues surrounding security with business owners, and they just want me to install some software or device to fix all their problems, especially when it comes to the human element of technology. Honestly, I also would love some type of artificial intelligence to solve all the problems out there for business. Unfortunately, there just isn't anything that can fix *us*—the biggest variable to computer and network security.

When deciding on how to design security systems inside a business, the first step is to decide exactly what you want to allow your employees to be able to do. I've seen a wide variety of responses to this, from "I don't want my employees to do anything online except to go into this one program they need for work" to "Let them do whatever they want. I completely trust my employees." Depending on your business, the correct answer is *in between*. Additionally, one of the reasons to partition your systems is that different departments will need access to different systems, especially when it comes to web access. Maybe you don't want all your employees to access social media websites, but certainly your marketing department will need access to be able to post to those same sites you don't want your employees accessing while on the job.

An acceptable use policy (AUP) is generally contained in your employee handbook but can be a separate document. It needs to define the behavior you expect your employees to follow while using your systems. Once it is defined, your technology professional can implement systems either to make sure it is followed or to notify you when it is not. It is up to management to make the decision on how strict the systems should be. When considering developing an AUP, here are some questions you can consider:

- Who in the company will oversee and enforce this policy?
- Do we allow employees to bring your own devices (BYOD) for access to corporate systems, or do we only allow access from company-owned devices?
- What different departments need different access to the Internet?
- Do we deploy a white list of websites that employees can access or a blacklist of sites or categories of sites that they cannot access (default to allowed or default to denied)?
- Do managers, owners, or other key individuals have unlimited access to the Internet?

- Can users install and update their own systems, or can only administrators make system changes?
- Will we potentially monitor employees' emails, computer usage, web usage, or any other technology usage on company time?
- Are employees allowed to use removable drives or online file sharing systems that allow for large exchange of information?
- Can our employees use company computer systems for any personal use, even when they are on breaks?
- What data requires encryption across the network? How will that data be stored, and how will the users be trained on identifying data that requires encryption?

If you have a human resources professional or outside firm, it would be important to bring them into the discussion because many of these decisions will impact human resources, and their input is essential. Remember, most technology problems are not really technology problems at all; they are human problems. Your users are the biggest security threat to your business, and controlling how they use your systems is a key element to protecting your network. You can find an example of an AUP on my website at www.nathanwhittacre.com/resources.

Employee Training

Employees are your number one vulnerability. They are the easiest gateway for hackers to enter your organization. We'll go into depth with this topic in chapter 6 ("Social Engineering"). Suffice it to say that employees, even well-meaning ones, can do significant harm to your company. Helping them identify activities that can harm the organization will be your best line of defense against outside hackers. It is also important to train them on the use of your systems, especially general computer use, email functionality, line-of-business applications, and your company's policies and procedures.

Suffice it to say that employees, even well-meaning ones, can do significant harm to your company.

I've always been surprised how many employees of our clients are unfamiliar with technology, even though it is integral to the company's operations. Organizations that invest in their employees through continual training improve productivity. If your employees struggle with computer systems, no matter their job function, they will not perform as well as highly skilled employees. The old saying applies here. One manager said, "If I train my employees, they will leave the company, and that money will be wasted." The other manager replied, "What happens if you don't train your employees and they stay?" Investing in your employees with continual training in technology will improve their lives, and, in turn, they will help improve your company.

KEY TOPICS YOU SHOULD INCORPORATE IN YOUR TRAINING

- Office applications such as Microsoft Word, Excel, Outlook, Teams, and PowerPoint (if used in your organization) or other general office systems
- Line-of-business applications—always pay for training with the vendors and have the option for continual training

- Internal processes and procedures that differentiate you from your competition
- Identification of social engineering techniques from hackers
- Identification of employee theft—both of time and of materials
- Protection of the company's intellectual property

In addition to the training, I recommend that you implement a testing protocol, especially for spam email detection. We'll cover this type of testing in the next chapter. Remember that any training for your employees will add significantly to your layers of protection against both internal and external threats to your company.

Next-Generation Antivirus Solutions

Since the early days when I was working on old computer systems (think black-and-white DOS computer screens), antivirus systems have been the staple of protection of endpoints (desktops, laptops, and other user devices). Antivirus systems have traditionally had a list of known malware and compare the files and software on the computer with that list. This means that the company that provides the antivirus solutions has to have seen and analyzed that malware inside their labs and added it to their list. They then push that updated list out to all their customers throughout the world so that the software knows about the latest malware out there. This is similar to how a vaccine manufacturer creates a medicine to fight off an infection from the flu every year.

Unfortunately, malware creators can just change a few aspects of their virus, and the antivirus creators will have to publish a new list. This often allowed the bad guys to have a few hours or days when no one knew about a new virus, and they could infect millions of computers before there was a "cure." The antivirus creators started incorporating heuristics inside their software, which allowed it to detect software that acted like a known virus and block that also. This did a great job at protecting against similar threats to older ones and prevented many attacks in that short window between the virus being released and the antivirus companies pushing out the cure to all their customers.

Modern endpoint detection and response (EDR) systems, or next-generation antivirus, monitor the behavior of the computer systems against the normal software behavior that the EDR would expect. Then it collectively compares that behavior against other systems running on the same network or across all similar systems globally. If the behavior is not what is normally expected, it flags the software, website, or other user interaction as potentially an issue and notifies the network or security administrator. This way new threats are detected much faster, and attacks are thwarted before they even start. Additionally, it is monitoring user behavior and could potentially prevent a user from being able to do things that are against the AUP or other expected behavior. It can also automatically isolate or shut down potentially infected computers before they can cause issues across the entire network.

Zero-Trust Framework

Historically, the security framework of computer systems has been a "trust but verify" mentality. Once a program or user gains access to a system, it is generally allowed to behave across the entire computer infrastructure without additional verification, so long as it (the user or program) behaves normally.

Think of it this way: if you are going to go to a party that requires an invitation to get in, once you pass through security to get into the party and the security guards verify you, you are generally free to roam around the party so long as you don't start jumping on the tables or throwing food at the other guests—unless it is that kind of party, then just carry on and have fun! As long as your behavior doesn't raise any red flags against the normal expected behavior of the other guests, you can enjoy the party.

Because of increasing attacks on computer systems, a new framework has been developed in the last few years. The basic premise of the new zero-trust framework is exactly what the title states: all users and software have no access until they are explicitly verified at each step along the way. So when you get in line to get some food, your credentials are verified. When you talk to anyone at the party, they also ask you for credentials. When you go to the bathroom, you guessed it, credentials again. This way, if a party guest changes in any way because maybe they had too much to drink after a few hours, they can be stopped before spilling a glass of wine all over the party host's white dress.

Zero trust on computer systems means that access is default denied, not just at the perimeter of the network but also at each step along the way. This idea could potentially cause users some pain because they may not be able to do exactly what they want all the time. On the other hand, the extra protection that this offers allows security professionals to protect against attacks that they don't know about. Additionally, it can also prevent attacks from hackers or even employees who attempt to gain access to systems that they normally do not access. It reduces the burden to constantly update against the unknown, such as new malware or employees who are no longer part of the organization that had their credentials stolen, because it treats everything as unknown until proven otherwise.

Zero-trust implementation is advanced security for your systems. You shouldn't think about implementing it until you have other more basic things in place. I'm bringing it up because once implemented, it can give you even greater peace of mind because you're protecting yourself from the things you don't know about. Donald Rumsfeld, the former secretary of defense for the United States, once said, "There are known knowns; there are things we know we know. We also know there are known unknowns; that is to say we know there are some things we do not know. But there are also unknown unknowns—the ones we don't know we don't know" (Rumsfeld 2021).[3] Zero-trust framework is to protect you from the unknown unknowns.

Internal Cybersecurity Checkup

Give yourself a score of 1 to 3 in each area. Add them up to see how you did overall in this area. Focus on just one area to improve upon for now.

3 Donald Rumsfeld, Department of Defense Archives, January 13, 2021, http://archive.defense.gov/Transcripts/Transcript.aspx?TranscriptID=2636.

AREA OF CONCERN	SCORE
I have identified all the digital assets that I want to protect across my organization.	
My human resources and administrative departments actively notify network administrators when employees or vendors leave the organization or no longer need access to the network. The list of users and vendors is reviewed periodically by management.	
I have a current AUP in my organization that is consistently updated and agreed to by all employees and a system to enforce compliance with it.	
My computer and network systems are partitioned in such a way that users and guests only have access to the areas of the network that I have explicitly allowed.	
I have implemented an EDR system on all my endpoints, my network administrators know if any EDR system goes offline or is out of date, and they actively monitor and correct problems on the systems.	
TOTAL **TOTAL POSSIBLE 15**	
NOTES	

Internal Cybersecurity Key Points

- The biggest threat to an organization's security is internal—your employees and their behavior toward technology is your biggest risk.
- Your company's management must be involved in defining the company's security policy and be the best example of good behavior.
- No company is too small to implement good security. The threats of loss to a small company, even an individual entrepreneur, are too great today to ignore.
- Good security starts with a good policy.
- Security protection requires layers. There is no one system to protect against all threats.
- What worked for security yesterday will not work today or tomorrow. All policies, systems, and implementations must be constantly monitored and updated.
- Better security is implemented by default to deny rather than to default to allow.

CHAPTER 4

EXTERNAL CYBERSECURITY PART 1: ENDPOINT THREATS FROM OUTSIDE THE ORGANIZATION

One of the most difficult hikes that I completed was the High Sierra Trail in 2020. It isn't hard to forget how hard 2020 was, with COVID-19 disrupting everyone's lives and feeling trapped in quarantine. I was really looking forward to hiking that year and getting away from the daily struggles of life for just one week. The headwinds were still difficult, though, because of potential cancellations due to the virus and the forest service's uncertainty about releasing permits. I still forged ahead and obtained permits to hike a seventy-mile stretch from Crescent Meadows in Sequoia and Kings National Park to the top of Mount Whitney. It is a beautiful and difficult hike and is regarded as the little sister to the John Muir Trail.

Adding to the difficulty of the year, the 2019–2020 winter was very dry. The Sierra Nevada only received a fraction of the average amount of snowfall. This poses two problems for hikers. First, water supply through the hike will not be quite as plentiful as a normal year. And second, the danger of large-scale fires increases dramatically. By the end of the 2020 fire season, more than 4.2 million acres burned, which is equivalent to the entire area of Los Angeles, Orange, Santa Clara, and Santa Cruz counties combined. (California's 2020 Fire Siege: Wildfires by the Numbers 2022).[4] Before my planned hike, the Sequoia Complex fire was spreading through the Sequoia National Forest, the Inyo National Forest, and the Golden Trout Wilderness. All of these areas were still to the south of the trail, but it meant possible smoke on the trail and evacuations if the fire continued northward.

4 Julie Cart, "California's 2020 Fire Siege: Wildfires by the Numbers," Cal Matters, July 29, 2021, https://calmatters.org/environment/2021/07/california-fires-2020/.

Even with these outside dangers, my hiking partner and I decided to proceed with the hike. We felt that the danger from the hike was acceptable, and we could proceed with it. The first two days of the hike were not terrible with smoke. The smoke seemed more in the distance, blocking some views but was reasonable. We started the hike on September 5, 2020. Unbeknownst to us, the day before, the Creek Fire started in the Shaver Lake area, which is to the north of where we started the hike. The fire quickly spread, burning ferociously throughout Sequoia and Kings National Park. Because of the fire danger, the park was closed a few days later and necessitated the rescue by helicopter of hundreds of hikers. By midhike, the smoke was terrible, with ash floating through the air in the morning and the sky a burned orange color throughout the day. The air made it difficult to breathe, which added to the dangers of continuing. I contacted home and received information about the closure of national parks and the growing fire danger. Our only course of action was to press forward as quickly as possible. We joined a few other hikers to increase our pace and exit the trail as quickly as possible. Throughout the rest of the hike, rangers posted notices that hikers needed to exit the trail as quickly as possible, which for us was to complete the hike.

Luckily, as we continued the hike, the smoke got better, and we were able to start picking up the pace. Even though we felt the urgency to get off the trail, it was an enjoyable hike. It pushed all our capabilities to get through the next few days. The views from the top of Mount Whitney were incredible, and at 14,505 feet I could see the smoke from the fires raging in the distance. I went back to hike the same trail with my son the following year, and several of the areas that I had hiked through the previous year were severely burned. It was a clear reminder to the dangers that were present. Within about a week of hiking through the Kern River area, the Sequoia Complex fire burned thousands of acres of beautiful forest that stood majestically while I hiked in it. I felt an eerie sense that I could have been trapped and seeking a rescue just a few short days after I had exited the danger areas.

There are many other dangers while hiking; most of them are external. These include weather, which my son and I encountered the following year with a sudden thunderstorm hitting us while near the top of Mount Whitney and drenching us with hail, snow, rain, and lightning. Others are animals, such as black bears looking for and stealing food or marmots chewing through tents and equipment to get leftover snacks forgotten in bags. There are creek crossings that are extremely dangerous to cross during a high snow year. Finally, dangers from other hikers, mostly those who are unprepared for the hike, who will be seeking assistance from others to get them through. No matter how well a hike is planned, there are always unexpected dangers that have to be mitigated because they can't always be avoided.

Cybersecurity planning is much like planning for a challenging hike. There are constant dangers surrounding your business. Many of them are well-known and defined. Those are preventable and generally easily mitigated. Other threats are unknown or unexpected. For those, it is important to have contingency plans in place in case disaster still strikes, no matter your preparation. Having a good team in place to support you along the journey is essential. Training, communication, and a strong support network are all essential to completing your digital journey to its end.

On the next page is a list of key definitions that you can refer back to as you read this chapter.

Key Definitions

- **Spoofing:** Making a connection (or call) appear to come from a different origin than from where it actually originated.
- **Malware:** Software that is intended to produce an undesirable result on a machine or network, often resulting in the destruction of data, a network breach, or further intrusion.
- **Internet of Things (IoT):** Any device that connects to the Internet to provide some type of functionality to the end user. These are devices such as smart watches, home fitness equipment, manufacturing devices, or data collection points across a corporate network.
- **Firmware:** Software that runs directly on a device that provides the functionality of that device. It can be considered the operating system of a device, such as a router, small electronics devices, or other IoT device.
- **Router:** A device that connects two or more networks together and allows the devices on those networks to communicate with each other.
- **Firewall:** A piece of software that either blocks or allows inbound and outbound connections to a machine or network.
- **Penetration/breach/intrusion:** When an individual gains unauthorized access to a network.

- **Brute-force attack:** When a hacker attempts to guess an access combination, usually by attempting common usernames with common or random passwords.
- **Black hat hacker:** A group or individual that attempts to penetrate a network or infrastructure without the knowledge or permission of the owner.
- **White hat hacker:** A group or individual that is hired by an organization to attempt to breach a network or infrastructure using the same methods as a black hat hacker to find vulnerabilities in the system.
- **Red team:** A group that acts in the role of the hackers to consistently find vulnerabilities in the corporate technology system.
- **Botnet:** A group of computer systems that are controlled by a hacker to search the Internet for victims, launch attacks, or provide computing power for illegal activities.
- **Endpoint detection and response (EDR):** Monitors and detects suspicious behavior and activities at network endpoints.

Detecting and Preventing External Threats to Endpoints

The majority of threats to a computer system or network will come in from the outside. I'll generally use the term "hacker" from now on to identify someone who you don't want to get access to your computers or data. This could be a black hat (someone you don't know and wouldn't want to access your network) or a white hat (someone you hired to act as a hacker and see where your vulnerabilities are). Either way, you don't want a hacker to enter your system, steal your data, use your computers for further hacking, or hold your network for ransom. Going back to the layers of security that I discussed in chapter 3, this chapter will focus on the perimeter and network portions of the layers of security. Aside from the human layer of security, these are your two biggest defense areas, which are constantly under attack.

For years, I've managed servers that are connected to the Internet that provide services such as web hosting, email, VoIP phone systems, and many other services. I can't say I'd do it for entertainment because you'd think I'm a complete nerd, but I would sometimes watch the server logs to see how often the devices were being attacked. Hackers would attempt to use a brute-force attack constantly by guessing common usernames, such as admin, administrator, root or a combination of people's names, or potential usernames, along with common passwords (e.g., password123 or qwer1234) or random passwords. Without an active intrusion prevention system, the hacker could go on for hours or days guessing the combinations. Do you wonder how long it would take a hacker to guess your password?

The next chart may scare you a little bit.

NONCOMPLEX PASSWORD LENGTH (LETTERS ONLY)	TIME TO BRUTE-FORCE GUESS (CRACK)
6 characters	Instant
8 characters	22 minutes
10 characters	1 month
12 characters	300 years
14 characters	800,000 years
16 characters	2,000,000,000 years

COMPLEX PASSWORD LENGTH (LETTERS ONLY)	TIME TO BRUTE-FORCE GUESS (CRACK)
6 characters	5 seconds
8 characters	8 hours
10 characters	5 years
12 characters	34,000 years
14 characters	200,000,000 years
16 characters	1,000,000,000,000 years

One caveat to this is that you need to use different passwords on different sites because if a hacker got your sixteen-character complex password from a website that they got access to, they would be able to use it instantly. We'll get into password management in chapter 12. Suffice it to say, many of your users have short passwords that are easy for a hacker to crack.

Going back to my story of watching a hacker attempt to gain access into the systems, it was always interesting to me how easy it was for them to persist in their efforts. This is because hackers build complex networks of computer systems that they have already hacked and use those to hack into other systems. These are called *botnets* and are a very powerful tool. These botnets are searching across the Internet for systems that have vulnerabilities, have weak security, or may just have potential to be accessed. Even though you can't see it, these types of attacks are happening against your network all the time, 24/7/365, and are unrelenting. The hackers just need to be right one time to get into your network. You have to be right 100 percent of the time to prevent the attack.

Intrusion Prevention

Other than the human layer, the perimeter of a corporate network is the largest layer to protect against. This is the part of the network infrastructure that is connected to any external network. In the simplest terms, this is the *company's connection to the Internet.* Not too long ago, when most companies were operating out of one location, with one Internet connection and no remote access to the network, this was much simpler to maintain. It generally consisted of one router/firewall protecting the entire network. In the last few years, the perimeter has become much larger, with employees now working remotely, companies expanding to multiple locations, and moving data and services to cloud services, such as Amazon Web Services or Azure. Each site, whether it is an employee working at home, at a Starbucks, from a hotel, or from the office, or the services running in the cloud, needs to be protected. Each location is a potential vulnerability.

A common misconception I hear is that the cloud is more secure than hosting data in their office. Another one is that no one really wants "our data," so it isn't important to protect it. Many people don't even give thought to how to protect against company computers running from a home or other insecure network connecting to corporate data. Preventing intrusion from multiple locations is much more difficult than when there was one entry point into a corporate network. It gives the attacker more possible points of entry into a network, many of them with limited or no protection at all. Even worse, many of the devices are running on networks that have other computers and devices on them that are also unprotected—thank your teenage children who are running all kinds of devices at home—that could be actively attacking your employees' computers while they are trying to have a Zoom meeting or while writing that important customer proposal. Because of this, we must look at intrusion prevention on both the network perimeter and the device level.

PERIMETER INTRUSION PREVENTION

When I first started connecting computers to the Internet, it was usually through dial-up connections. Technically, the computer was connecting directly onto the Internet, receiving a public IP address from the Internet service provider (ISP), such as America Online (AOL). Remember getting those floppy disks in the mail with your free five hundred minutes? Most computers didn't have a firewall built in at first, so other computers on the Internet could potentially connect to any other computer connected to the Internet. Attackers figured this out and were able to access computers, steal data, and spread viruses. Most computers today have personal firewalls built into them to prevent direct attacks against them when they are connected to the Internet. I'll get into how to properly set those up in a later section. For now, remember that having a personal firewall (usually built into the operating system or part of the antivirus system on your computer) is essential for all devices.

When cable modems, DSL, and T1 connections started becoming available, the need for another device became important to allow all the computers on a business or home network to connect through one shared Internet connection. This device is called a router, which allows two networks to be able to communicate with each other. Under what is called Internet Protocol version 4, or IPv4, each device that is connected to a network needs an IP address. These can either be publicly routable IP addresses (ones that any computer connected to the Internet can access) or private IP addresses (ones that only computers on a local network can access). Think of public IP addresses like your street address. If someone tells you to go to 123 Main Street, Anytown, USA, anyone in the world can find your location. That is your network's public IP address. This is usually four sets of numbers from 0 to 256 separated by dots. For example, 8.8.8.8 (this happens to be a server hosted by Google) or 121.244.45.78 (I have no idea who this belongs to, so don't show up there with your mom's favorite casserole).

On the other hand, your business or home network generally uses a private address. When IPv4 was designed, they set aside several blocks of IP addresses that can be used for devices on local area networks (LANs). These are 10.0.0.0–10.255.255.255, 172.16.0.0–172.31.255.255, and 192.168.0.0–192.168.255.255. The router that connects two networks together, whether it is public to public or public to private, is to provide a gateway between the two networks. Routers that connect a public-to-public network have an easier job because there is no translation necessary between the two networks, as both sides can see each other's IP address schemes. It just needs the router to let it through the intersection.

Public-to-private routers have a bit more work because they must translate between the private network addresses to usually a single public IP address and keep track of all the connections. Initially, routers didn't have much memory and processing power, so they ran as stateless routers. This meant that when a computer on the private network would request access to a website or some other server on the Internet, they would open a port or temporary door between the outside Internet and the local network but just leave that door open for anyone to come back in. It would only close the door when no data had come through it for some time or when the computer on the local network would ask the router to close it. A stateless router would just allow anything to come into the local computer, even a hacker showing up uninvited with his mom's casserole—except it is filled with the latest computer viruses.

To prevent this ill-advised, open-door policy, routers started to implement stateful firewalls in them. This type of firewall tracks each individual connection from an internal device to an external server in a table inside the router's memory. Whenever the external server sends data back to the router to forward onto the internal computer, it checks that connection in the table to make sure the connection is valid. This required additional processing power and memory to be able to store information about all the different translations that were being requested. The user doesn't see it, but opening just one web page can create thousands of requests to many different web servers on the Internet. Every image, drop-down, and interactive element and those wonderful ads that show up on a website usually come from different places. The router must keep track of all of them and ensure that the communication is only occurring through that open door between the server that the private computer initially contacted and the internal device. Everything else is blocked by the active firewall. All this work can slow down the Internet connection, so a helpful tip for picking out routers today is that they are generally rated to a maximum speed of connection that they can handle. With many business and home connections approaching gigabit speeds and beyond, your router can limit the actual speed you can get on your Internet connection if not purchased properly.

A brief side note here: I will generally use the terms "routers" and "firewalls" interchangeably because the functions of routing and firewalling are commonly built into one device today. For clarity, routing is the function of connecting two or more networks together and allowing them to talk to each other. A firewall protects that communication between the networks, allowing only the data that is supposed to flow between them into the device.

Unfortunately, hackers didn't just walk away from attacks on the perimeter of the network because you installed a smart door to control who can go in and out of your high-tech home. Hackers figured out many ways to get past stateful firewalls. One of the most common ways is to run active port scans against a router. This is similar to walking around a house checking for open doors or windows. Most routers will allow this activity to go on unimpeded forever. An attacker can keep checking, over and over again, until it finds a vulnerability and gets into the network. Often, these vulnerabilities come from the firmware in the router or a device that is misconfigured. When a network administrator programs a router, ports can be opened to allow for direct communication from the public

Most routers will allow this activity to go on unimpeded forever. An attacker can keep checking, over and over again, until it finds a vulnerability and gets into the network.

network into private servers. This is useful if the network is running a server, such as an email server, line-of-business software, or web server. Other times, remote users want access into their computers at the office, so a network administrator will open up ports on the firewall to allow for communication to those computers directly. Each time a port is opened, it leaves a door open for a hacker to have direct access to a computer or server on the private network. That door may need to be open to allow for the services that the server provides, but it also leaves the network more vulnerable to an attack.

To provide better protection against these more advanced attacks, advanced firewalls with active intrusion prevention services (IPS) were created. Rather than just keeping track of connections going in and out of the open ports on the router, an active firewall inspects all the data that goes in and out of the network. It is like setting a security officer at each door that goes in and out of your house. Think of it like the TSA for your business network, except the good ones will do the full-body scan, the bomb screen, and all the intrusive pat downs that no one ever wants to go through before getting on the plane. An active IPS will look at each data packet going in and out of the network, comparing them against known attacks and viruses. They will also monitor port scanning and actively block remote computers that have "knocked" too many times. They also have an active list of IP addresses that are *known* bad addresses—ones that have been shown recently to have hacker-type behavior—and not allow connections from those addresses.

Another advantage of having an active IPS is that it can also prevent connections from the internal network to locations you don't want the users to go to. At my company a common request is to enforce the AUP through technology. We do this by creating content filtering on the firewall. The IPS can actively block connections to sites that contain malware and viruses, pornographic websites, social media, video streaming, or other sites that management doesn't want employees accessing. This is also a great way to make sure employees are actively doing their work rather than spending time looking for their next job while they waste the company's time. An active IPS can also interface with your other security systems, such as your SIEM, which I will describe more later, to collect an overall picture of the security threat against your network.

One important takeaway from this section is that setting up a firewall correctly is not simple work. It takes a skilled network administrator to make sure all the settings are correct. I often go into new clients and find a router that is out of date, has no active subscription for IPS, or has ports that were left open to computers that are no longer active. Buying an advanced firewall but not configuring it correctly or maintaining it is just as bad as not having one at all. You should review the settings of the firewall at least annually to make sure everything is current and operating as expected. Most of them have annual subscriptions that need to be kept current, which is a perfect time to update the firmware, check all the settings, and ensure that it is configured correctly.

Endpoint Intrusion Prevention

A good IPS firewall will only protect devices that are connected to the Internet through it. What about the devices that are being used remotely—at home, in the coffee shop, in a hotel, or anywhere that isn't protected by the business infrastructure? Mobile devices are more difficult to protect than computers that are just kept in the office all the time. On top of the increased difficulty, it is more important to protect them because they are living in the wild and can bring back the remnants of what they are exposed to when reconnected back into the office.

One of the major concerns about mobile devices is the data that they contain. In 2006 a Department of Veterans Affairs laptop containing approximately twenty-six million living "veterans' social security numbers, full names, birth dates, service numbers, and combined degree of disability" was stolen.[5] While the contractor should not have had that much data stored unencrypted on a removable hard drive, it doesn't negate the fact that the person did, and the data was stolen. Are your employees storing data from the company on their laptops, phones, tablets, etc. that can possibly be stolen? How do you know that they are not? In the case of the VA, the department was not aware that the contractor had that data, and they also were not aware that the data was being stored unencrypted. Both of those actions were in violation of the policies of the VA.

Full-Drive Encryption

You can take a first step to protecting your company's data by ensuring that all mobile devices have full-device encryption enabled by default. Most Microsoft Windows laptops with the Professional or Enterprise version of the operating system have a solution called BitLocker built into it. The laptop should also have a Trusted Platform Module (TPM) installed on it. This enables the operating system to be able to encrypt and decrypt the data in hardware rather than software, which is much faster. To check to see if you can enable BitLocker on Windows, follow these steps:

5 Department of Veterans Affairs Office of Inspector General, "Review of Issues Related to the Loss of VA Information Involving the Identity of Millions of Veterans," July 2006, accessed April 3, 2023, https://www.va.gov/oig/pubs/VAOIG-06-02238-163.pdf.

1. Right-click on the Windows menu in the lower left corner of your screen.
2. Select System. Look for the Windows specifications section. It should say Pro or Enterprise. If it says Home, you will not be able to easily get full-drive encryption.
3. To check if you have a TPM module, right-click again on the Windows menu and select Device Manager.
4. You should have a section that shows Security devices. Click on the down arrow next to that, and it should show TPM.
5. Finally, you can check to see if you have BitLocker enabled on the laptop by right-clicking on the Windows menu and selecting Settings.
6. Select Update & Security and then select Device encryption. It should show that Device encryption is on.

If you are running macOS, FileVault 2 is available in OS X Lion or later. When FileVault is turned on, your Mac always requires that you log in with your account password. You can also enable full-disk encryption by following these steps:

1. Choose Apple menu () > System Preferences, then click Security & Privacy.
2. Click the FileVault tab.
3. Click , then enter an administrator name and password.
4. Click Turn On FileVault.

If you do not have any of these options on your mobile devices, you will need to either replace the device with a new laptop that has these options or purchase a third-party tool to perform the full-drive encryption. I highly recommend replacing the device, especially if you are running Windows Home edition for your business laptop. The Professional and Enterprise editions contain other security mechanisms that are essential for business operations. They call it Home edition for a reason. It is designed for personal use only.

DNS Filtering

When mobile devices are off the company's network and using another Internet connection, they often are using another DNS, which is a server that translates a website name into the IP address that the computer can use to contact that website or service. There are two issues with using other DNS servers. First, DNS queries can be used by a hotel, coffee shop, or other service provider to track the usage of the Internet at their location and potentially redirect web queries to another site. Have you ever logged into a hotel's website and been directed toward a "splash page" or a site that makes you put in your room number and name to be able to use the hotel's Internet service? That happened because the router that you connected to with your laptop redirected you from your home page to the hotel's website using DNS redirection. Obviously, this is a nonmalicious use of DNS redirection, but the hotel could also use it to redirect your query to go to one restaurant over another if you did a search. They can also monitor all the websites you are going to and use that information to show potential advertisers what sites their guests are visiting. Do you think they can't do this? They can! You agreed to it in the terms of use agreement that you clicked on when you logged into the Internet. You gave them permission to collect that information on you.

Another more devious way of using DNS redirection is DNS hijacking or DNS poisoning. This is used by a malicious attacker to redirect your Internet usage to their websites rather than the ones that you expect to use. For example, if a hacker wanted to get your email credentials, they could set up a website that looks just like Microsoft Office 365 but with a different IP address. If they are able to hijack your DNS request, it would send you to their *Evil 365* website rather than Office 365. You would enter your credentials and even your two-factor authentication, which the hacker would immediately use to log in to Microsoft's real Office 365 website, and they would have complete access to your email. This is one of the reasons why *you should never connect to a Wi-Fi service that you are not familiar with.* I generally recommend connecting to your cell phone's mobile hotspot rather than using random Wi-Fi services in restaurants or retail locations. A hacker can easily set up a Wi-Fi service in the parking lot and name it something similar to the store. They could get people to connect to the service and hack into their accounts by collecting information while just sitting in their car.

One way to protect yourself while roaming is to use a DNS protection service. Software can run on your computer and act as a secure DNS server. That software then connects to the company's managed DNS service to securely make the DNS queries. The network administrator can then enforce the company's AUP on Internet usage by also blocking access to websites that management does not want the employees going to, such as social media, pornography, or other non-business-related sites. Generally, the company that provides the DNS filtering software also provides a list of websites that are known as bad sites, those sites that hackers set up and use to hijack DNS or that contain viruses and other malicious information. Some examples of these services are Cisco

Umbrella, DNSFilter, and OpenDNS (which is now part of Cisco). For home use, I personally like OpenDNS because they have a kid-friendly service that automatically filters out sites you would not want your children going to, and it is free for personal use.

Mobile VPN Service

Another popular mobile protection option is to set up a virtual private network (VPN) service. A VPN is basically a protected tunnel through the Internet. Think of a VPN like a pipe running under the ocean. Whatever you send through the pipe is still going through the ocean, but it is protected against the water and other potential contaminants that are in the ocean. It should get safely to the other side without interacting with the rest of the things that are in the ocean. A VPN is similar, where each side encrypts the data that it is transmitting back and forth. When your laptop sends data to the VPN provider, which could be the corporate network or a personal VPN service, it first encrypts the data, then sends it out to the Internet. If someone was able to intercept that data, they couldn't do anything with it because they don't have the right key to unlock the data. Once the other side receives the encrypted information, it can safely process it. Most personal VPN service providers then run the data that is being sent back and forth through antivirus, intrusion prevention, deep packet inspection, DNS filtering, and all the other protections that the corporate network administrator would want for you if your laptop was in the office. If you do plan on using a personal VPN service, you should run it by your IT professional. Many of the personal VPN providers are in

If you do plan on using a personal VPN service, you should run it by your IT professional.

different countries and adhere to different laws. Because all communication from your mobile device to the Internet is through that VPN service, they could also be doing DNS hijacking or other malicious things with your connection. I personally feel that a bad VPN service provider is worse than not using one at all. A good VPN service is extra protection as long as you are doing everything else in a safe way.

Mobile Phone Protection

Cellular phones are also a target for hackers to penetrate the corporate network. These devices are minicomputers and generally have direct access to email and other servers. If your company is using systems such as Microsoft Office 365 or Google Apps, especially with single sign-on (SSO), there is a potential opportunity for a hacker to get into a device and then use the SSO privileges to access the corporate email and other servers.

There are a few things that should be done to protect mobile devices.

1. Ensure that when a mobile device is connected to corporate email services, remote wipe capabilities is enabled. In the event that the device is lost or stolen, the email administrator can remotely erase all corporate data from the device and destroy the link between the mobile device and the corporate network.
2. Enable device passcodes in iOS (iPhones, iPads, etc.), which will automatically enable full-device encryption. Your system administrator can enable the option in Microsoft 365 to require the devices to have a passcode.
3. Because there are many versions of Android phones and devices, the settings are somewhat different between devices.

Generally, encryption settings are enabled if you create a device passcode and screen lock. To verify, there are generally options under Security and Encryption that allow you to enable device encryption. Similar to iOS devices, a system administrator can enforce device passcodes and remote wipe capabilities.

4. Do not download and install third-party applications that are not in the App Store or Google Play Store. The Apple App Store is much more closed and less likely to have apps that contain malware, but it is still possible. Android phones are more susceptible to malicious apps because they are more open, and you can install apps from unverified third parties. It is important to train your employees to not install apps on their devices that could potentially be harmful.
5. Biometric passcodes are generally more secure than PINs or short passcodes. Requiring a biometric passcode (Face Recognition or Fingerprint Scanning) is better protection on mobile devices.

In many small businesses, cellular and mobile phones are a BYOD model, where the employees use their personal devices to access company data. Because of this, it is nearly impossible to limit what is being installed on those devices besides corporate data. Having protections against those mobile devices in the corporate infrastructure is where system administrators can provide the best protection for your company.

External CyberSecurity Checkup

Give yourself a score of 1 to 3 in each area. Add them up to see how you did overall in this area. Focus on just one area to improve upon for now.

AREA OF CONCERN	SCORE
Complex and long passwords are required across all devices and systems in my corporate infrastructure.	
The company has a modern firewall with active intrusion protection that is monitored and up to date.	
DNS filtering is enabled on all corporate devices.	
Full-device encryption is enabled on all mobile devices and any computer that contains sensitive data.	
Mobile device passcodes and encryption are enforced by my email systems.	
TOTAL **TOTAL POSSIBLE 15**	
NOTES	

External Cybersecurity Key Points

- Hackers are constantly attempting to find vulnerabilities in corporate networks.
- Beyond the human element, the perimeter of the network is the largest layer of needed protection.
- Firewalls have advanced over the years. It is essential to keep your firewall up to date and all security protections active.
- Mobile devices make it much harder to protect the perimeter of the network because it is constantly changing and growing.
- Mobile endpoint protection is just as essential to protecting the corporate network as investing in security systems inside the network.
- DNS filtering is a powerful protection for mobile devices because the AUP can be enforced even if the device isn't behind the corporate firewall.
- VPN technologies can help with additional protections for mobile users.
- Bring your own device (BYOD) policies make it more difficult to protect the company's network.
- Most modern cellular phones provide full-device encryption once a passcode is enabled.

CHAPTER 5

EXTERNAL CYBERSECURITY PART 2: INFRASTRUCTURE THREATS FROM OUTSIDE THE ORGANIZATION

It's useful to draw a parallel with the need to prepare for the dangers of hiking and the need to prepare for the dangers of external cybersecurity. Why? Because backpacking *can* be dangerous. There are many external factors that affect a hike, and weather is one of the most difficult external factors to predict and protect against. Part of the problem with hiking in the Sierra Nevada is timing. There is a short summer window to hike safely without snow gear. It would be wonderful if that window was completely predictable, but Mother Nature is nothing if not unpredictable. The amount of snowpack in the mountains dictates the timing of a hike. If the winter has a heavy snowpack, hiking in June and much of July is very dangerous, actually one of the most dangerous times to hike. As the snow begins to melt, it becomes unstable. It also fills many streams, creeks, and rivers that run through the mountains. Many of those streams develop under the snow and are not visible to the hikers. Crossing a snow-covered stream, whether visible or not, can be very dangerous. For example, the snow could be developing a "snow bridge" over the stream. The bridge may appear stable, but in the afternoons, when the temperatures rise, the snow becomes softer, and a hiker's weight could collapse the bridge, plunging the hiker into a fast-moving, ice-cold river of water. In 2017, a very high snow year, there were two deaths in July on the John Muir Trail from drowning.[6]

To exacerbate the challenges of planning for weather, when a hiker is planning a trip, the snow conditions are unknown. The lottery and permitting window for hiking in the Sierra Nevada begins in December, about six months before the actual trip. Backpackers become mini weathermen, trying to forecast what the snowpack will be in the summer based on snow that hasn't happened yet. Most of the snow comes during January through March, with the high point of the snow accumulation around April 1. It is very difficult to know exactly what the hiking season will bring when getting the permits for the summer. It would be easy to say just go in August, which is generally the safest month to go. Because of the limited permits, it is difficult to get an August date. Pushing the date into September is

---

6 Veronica Rocha, *Los Angeles Times*, July 26, 2017, https://www.latimes.com/local/lanow/la-me-ln-hiker-body-found-kings-river-20170726-story.html (accessed February 10, 2022).
Ernie Cowan, *Pacific Crest Trail forms tight network*, August 5, 2017. https://www.sandiegouniontribune.com/communities/north-county/sd-no-cowan-column-0805-story.html (accessed February 10, 2022).

also complicated. Snowstorms can come in late September. The days are also shorter, the nights are colder, and the storms can bring an accumulation of snow and ice in the mountain passes. In other words, timing the hike based on unknown weather conditions is very tricky.

I've found planning and preparing for a thru-hike stressful. Much of the timing is guesswork based on historical and current information. Things can change quickly with just a few storms in late winter, a quick defrost, or a sudden surge of summer storms. On top of that, with increasing issues from drought in California, summer wildfires are increasingly common and more dangerous. The threats are constant against a successful journey. One of the things that I've used to help me over the years is peer groups that are also hiking in the same area. There are several experts in these groups who have compiled the latest information on transportation, weather data, past reports from hikers, and current conditions. These experts have been teaching about hiking in the Sierras, writing books, and collecting data for many years. They not only answer basic questions for new members of the group but also collect current information and provide it in meaningful ways. I think that without these experts, many novice hikers would have far less success completing their trips.

I've used a hiking metaphor because much like the dangers that hiking can present, the complexities of navigating cybersecurity are also daunting, especially with the rapidly shifting nature of technology. There are new threats each day, and it is very difficult to predict what will happen in the future. Attempting to navigate these complexities on your own is very challenging. But having a group of people to assist you who are familiar with these dangers and are constantly up to date on the many facets of cybersecurity will help you navigate the digital world successfully.

Like the solo hiker who doesn't use any additional resources to plan their trip, many business owners I talk to go it alone on technology and security. They think that their IT guy can handle their cybersecurity and keep them safe from all the threats to their business. *There is no way a single person can be an expert in all areas of technology.* Most IT people are more like firemen, going from one problem to the next and solving them. I find that companies that have an internal IT person or rely on a single person often have more issues because the same technician who is installing printers and computers is also responsible for the protection of the entire network. They often neglect the preventative measures needed to protect the business because the urgent requests from users tend to take precedence over the steps they need to take to protect the entire network.

A team consisting of a chief technology officer, chief security officer, IT professionals, and technicians at different skill levels, as well as security professionals and technicians at different skill levels, all need to work together to build strategy, solve high-level technical tasks, and complete low-level help-desk service requests. All these positions can be outsourced to a trusted provider. If your business is too small to build a team of professionals internally, look at options to work with a professional firm that has all these areas covered so that you are fully protected.

Below is a list of key definitions that you can refer back to as you read this chapter.

Key Definitions

- **Spoofing:** Making a connection (or call) appear to come from a different origin than from where it actually originated.
- **Denial-of-service attack (DoS attack):** A hacker who has access to a big Internet connection attempts to flood another network with enough data to shut down all other services to it.
- **Distributed DoS attack (DDoS attack):** A hacker uses a botnet to attack another network and shut it down from multiple connections (sometimes in the millions).
- **Geo-redundancy:** Having servers hosted in multiple locations so that if one location (data center) goes offline, the services provided by the location that is offline can be transferred to another location that is online.
- **Patches:** Updates to software and firmware that correct bugs or security problems with the system.
- **Uptime:** Measurement of the amount of time that a service is available for normal operations. Usually measured as a percentage of a year that the service is functioning normally and generally excludes allowed maintenance periods.
- **Service-level agreements:** A service provider's guarantee that it will achieve a proposed uptime and the provided compensation for not achieving it.
- **Ransomware:** Malware that holds data or information hostage until a ransom payment is received. May also be used for blackmail by threatening to widely release said information.

- **Logging:** Collecting information on the function of an endpoint, server, or networking device.
- **Security information and event management (SIEM):** A system that collects the logs from all endpoints, servers, networking equipment, and endpoint detection and response (EDR) to provide analytics and potential responses to threats that may be entering the network.
- **Managed detection and response (MDR):** Works in conjunction with the EDR to provide real-time human insight into threats entering the system. Generally used by a SOC to monitor all devices on the network.
- **Security operations center (SOC):** A 24 hours a day, 365 days a year team of security analysts and engineers who monitor, detect, and respond to threats on the network.

Detecting and Preventing External Threats to Infrastructure

Just like planning for a hike when you can't foresee the weather, preparing for a cybersecurity attack on your network is also unpredictable. If hackers would just announce to you how they plan to get into your computer systems, you could place defenses exactly where they are needed. Even though there is no way to know how much snow Mother Nature is going to leave in the mountains half a year before hikers are planning their trip, there are ways for you to know where your vulnerabilities are and plan accordingly. You must plan for everything and be prepared for the worst possible scenario. Even with all the preparation, you also should have contingency plans in place, which we'll cover more in chapter 10.

Although it may be easier for hackers and insiders to use your endpoints (laptops, PCs, mobile phones, etc.) to access your network, the bigger prize is to gain access to server and network infrastructure. From the servers they can generally access all the company's data, gain access to all the computer endpoints, and often take time to plan their future attacks. It is also more efficient to launch other attacks if the hacker has access to all the company's critical infrastructure. In this chapter we'll discuss how a hacker can gain access to your company's servers and network infrastructure and ways to protect against these attacks.

Uptime

Before we get into the details about infrastructure protection, let's talk briefly about how providers measure the expected downtime of their systems. First and foremost, if an IT company or person tells you that their services will *always be working*, they are lying. There is no way for a company to guarantee 100 percent availability of their systems. Computer systems are vulnerable to required maintenance, hardware and software failures, outside attacks, extended power outages, network outages, and a number of other reasons why they can and will go offline. All of these things can be mitigated and reduced to a minimum with proper planning and design.

IT professionals measure the amount of time that systems are expected to be online in "uptime." This is usually measured by the percentage of the year that the systems are operational, excluding scheduled maintenance periods. Many companies will guarantee an uptime and provide credits back to the customer if their service falls below the guaranteed uptime.

Below are examples of uptime based on a year.

UPTIME GUARANTEE	EXPECTED DOWNTIME IN A YEAR
99.9%	8 hours 45 minutes
99.99%	53 minutes
99.995%	26 minutes
99.999%	5 minutes

In other words, if a service provider tells you that their service provides 99.99 percent uptime, which sounds great, they expect the service to be down unexpectedly about one hour per year. They can also be down for additional time for regular maintenance that has to be announced to you beforehand. For example, Amazon Web Services, a large cloud hosting provider that is part of Amazon and used by thousands of businesses, has a guaranteed uptime of between

99.0 percent and 99.999 percent depending on the service.[7] Before buying any service, it is important to review the service-level agreements (SLAs) and make sure that they fit with your business model.

Denial-of-Service Attacks

Think about a time when you were waiting for an important phone call. It's one of those calls that you really need to answer, and you don't want it to go to voice mail. You're waiting right by your phone so you don't miss it. Then the phone rings. It's not the person you expect it to be; it's a wrong number. The phone rings again, wrong number again. This happens over and over again for the next several hours. Your phone basically becomes unusable because of all these wrong numbers. You think to yourself, who gave out your number to all these people, and why are they calling you at the exact wrong time? By the time all the wrong numbers stop calling, you realize that your important call came in but went to voice mail because you happened to be on the phone with one of the wrong callers. You missed out on a great opportunity because someone gave out your number, and you were flooded with bad calls.

The same thing can happen on the Internet. Each Internet connection has an IP address and a maximum amount of data that it can handle at one time. Your phone technically has a capacity of two: the live call and a call on hold. If both of those are filled up, it goes to voice mail or is just rejected. If a bad actor wants to shut your company down, basically not allowing you onto the Internet, they can make enough requests or send enough

---

7 *AWS Service Level Agreements (SLAs)*, n.d. https://aws.amazon.com/legal/service-level-agreements/ (accessed March 11, 2022).

data to flood your router so it cannot make any more requests. This is called a denial-of-service or DoS attack.

It is relatively simple to stop a DoS attack. The advanced firewall (hopefully you have one) detects the flood of incoming data from a particular address and just stops responding to it. This is kind of like blocking a number on your phone. It will still take up some bandwidth and potentially partially fill up your incoming Internet pipe, but your connection should continue to work. Once the advanced firewall blocks the incoming connection, it will look to the hacker that you disconnected from the Internet connection, sending the hacker to permanent voice mail. If the DoS attack is bad enough, you may have to contact your ISP to get them to block the IP address the attack is coming from.

BOTNETS

We talked in the last chapter about one use of botnets, where the hacker is able to use hundreds or thousands of computers to scour the Internet for potential targets. Since the hacker controls these computers, he can also use them to launch attacks. These computers, which could be home or business computers, are fully under the control of the "bad guy." The owner of the computers has no idea that there is an extra piece of software running on them. Usually, when a computer is part of a botnet, it will act no differently than any other computer until the hacker activates it to do something nefarious. Then it will act slow or be completely unresponsive, sometimes only for a short period of time.

For example, one of the largest and longest-lasting botnets was Emotet. The malware that created the botnet spread through Microsoft Word files came through email attachments. The victim would open an email that looked like an invoice, shipping notice, or some other type of file that seemed very legitimate. It was difficult for antivirus systems to detect it because the malware was polymorphic, which means that it would change its code each time it was sent out. That way the detection software was unaware that it was an issue and would allow the file to open. Once the computer was infected, it joined the botnet and waited for instructions from the command-and-control systems of the botnet. The user who opened the file didn't notice any change in their computer, so these systems would go on undetected until they would be part of an attack. After about seven years, in 2021 the Emotet network was taken down by a joint international task force. It is still unknown how many systems were compromised, but you can look it up and see if your computer was part of the botnet.[8]

DISTRIBUTED DENIAL-OF-SERVICE ATTACKS

To overcome the limitations of attacking a network with a single computer, an attacker can use a botnet to take down entire computer systems. Even though the power of an infected individual PC or server isn't anywhere strong enough to take down a website, server, or entire computer infrastructure, millions of computers attacking all at the same time is extremely powerful. It is like sending over a cluster bomb composed of millions of nuclear weapons. These attacks have been used to take down large companies and disrupt communications. For example, in 2021 a DDoS attack on the telecommunications company Bandwidth took down VoIP phone service for many companies across the United States for several days. Even though Bandwidth had DDoS mitigation services in place, it was not sufficient to prevent the attack.

8 Eurojust, Press Release: World's most dangerous malware EMOTET disrupted through global action, January 2021, accessed February 21, 2022, https://www.eurojust.europa.eu/worlds-most-dangerous-malware-emotet-disrupted-through-global-action.

The company estimates that they lost over $12 million in revenue due to the attack.[9]

HOW DDOS ATTACKS AFFECT SMALL BUSINESSES

You may think that hackers will only want to come after large businesses with DDoS attacks. By now I think you know what I am going to say: no business is too small to be affected by this type of attack. Of course, an attacker can go after your small company and take down your Internet connection. But there are a few other ways that you can be adversely affected by a DDoS attack.

You may think that hackers will only want to come after large businesses with DDoS attacks. By now I think you know what I am going to say: no business is too small to be affected by this type of attack.

First, you could have computers in your network that are part of a botnet. Many companies have good Internet connections, computers that are faster than a home computer (think of your server sitting in the closet), and a well-maintained network. These are perfect conditions for a hacker to use your network to attack someone else. Once your computers are part of the hacker's botnet, they can use your computers to launch attacks on other companies. Your unprotected computers are now completely under the control of the hacker and can cause outages at other companies, hacking into other systems and causing disruptions

---

9 Jonathan Greig, *DDoS attack cost Bandwidth.com nearly $12 million,* November 8, 2021, https://www.zdnet.com/article/ddos-attack-cost-bandwidth-com-nearly-12-million/ (accessed February 21, 2022).

all over the Internet. Additionally, with remote workers, this may include computers that are in many different locations, thus amplifying the attack and making it more difficult for a security professional to shut down a DDoS attack against them. We'll discuss more shortly about EDR and MDR systems that can detect dormant malware on your systems and prevent them from activating.

Second, your online presence can be attacked. Most companies host their email and website with external hosting companies. These companies also host other companies' websites and email systems, so they are more likely to be attacked. It may also be a way for a hacker to compromise the server that is hosting your website and infect other computers that visit your site, thus increasing their botnet. It would be wise to discuss this with the company that hosts your website and email and find out more about how they mitigate against botnets and DDoS attacks.

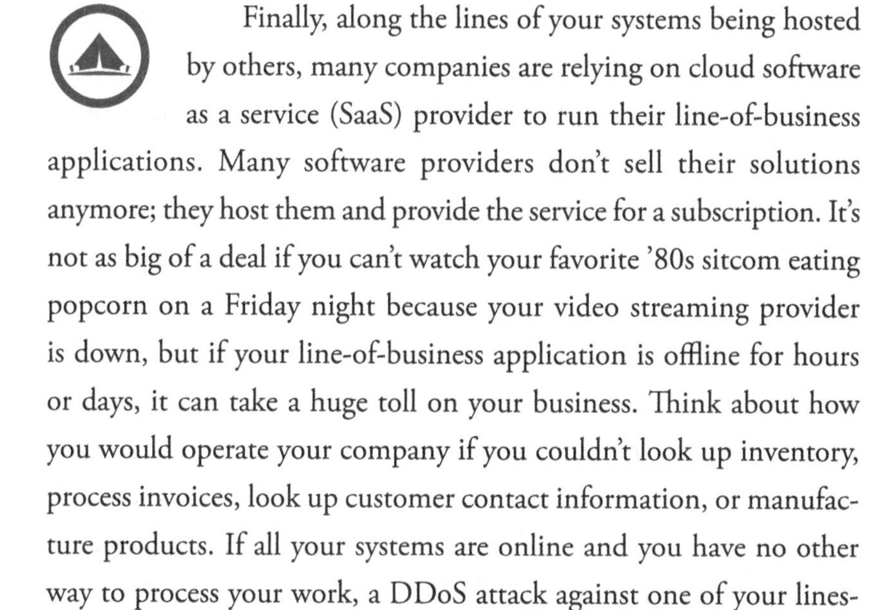

Finally, along the lines of your systems being hosted by others, many companies are relying on cloud software as a service (SaaS) provider to run their line-of-business applications. Many software providers don't sell their solutions anymore; they host them and provide the service for a subscription. It's not as big of a deal if you can't watch your favorite '80s sitcom eating popcorn on a Friday night because your video streaming provider is down, but if your line-of-business application is offline for hours or days, it can take a huge toll on your business. Think about how you would operate your company if you couldn't look up inventory, process invoices, look up customer contact information, or manufacture products. If all your systems are online and you have no other way to process your work, a DDoS attack against one of your lines-of-business application providers *can put you out of business.* Similar to a discussion with your website provider, a discussion with your SaaS provider about DDoS mitigation, geo-redundancy of data centers, and their disaster plans is essential to make sure you aren't a victim of another company's problems. Additionally, you should have backup methods to operate in the event that your SaaS provider is unavailable for several days, which has occurred for many of my clients.

Network Equipment Protection

Recently, I visited a prospect that was looking to sign up for services with my company. I generally walk around an office and look for equipment that has been installed by other providers. Like most prospects that I visit, I found old equipment that had previously been installed by another provider and was not being managed. The equipment was old, unsupported by the previous IT company, and running firmware that had security vulnerabilities. Just because a device doesn't look like a computer does not mean that it doesn't need updates and management. Routers, firewalls, switches, access points, printers, copiers, and anything else that plugs into the network require management and maintenance.

Security researchers are always looking for vulnerabilities in systems, especially devices that are supposed to protect the network. One prominent example of this from a few years ago was a vulnerability found in most Linksys routers.[10] A hacker could set up or exploit a website that would inject some code into the router that would reset its password and allow the hacker to log in to the router as an administrator. Once that was accomplished, the attacker could then install new firmware on the router that would allow further damaging exploitation of all computers behind the

10 Phil Purviance, *Superevr*, April 05, 2013, https://superevr.com/ (accessed March 03, 2022).

router. Millions of Linksys routers were sold and were in operation when this vulnerability was found. Linksys did release a patch to the system. Most users of their products, especially home users, would have had no idea how to update the firmware on their devices.

Some devices now come with options to have the devices automatically update on a schedule. I would recommend this for most personal routers and switches. For business devices, these should be monitored and updated manually. Firmware updates can cause other unintended issues, such as the device not coming back online after a reboot to install the firmware or changing a configuration setting that causes other problems on the network. Additionally, a backup should be made of the device's configuration before *any* updates so that it can be reverted to previous firmware and settings if something does go wrong.

If you remember all the way back in chapter 2, we discussed *setting a hardware life cycle for devices.* Most network equipment has a life span at the end of which the manufacturer stops releasing updates and patches for the device. If, for example, Linksys received notice of that vulnerability to a device that was at the end of its life, Linksys would have had no obligation to provide patches or updates to it. The recommendation from the manufacturer would be to replace the device with a modern router.

Additionally, most advanced routers and firewalls require a subscription to maintain their security services. These should be kept up to date. I generally recommend subscribing to them annually, which gives a good opportunity for the system administrators to log in to the device, update the firmware, and make sure all the settings on the device are still valid and functioning as expected. Of course, the administrator should be doing that more often, but at least once a year is better than never. There's nothing like a hard deadline to get something done. Also, an annual subscription is a good chance to review to see if the device is still valid or will be reaching end of life in the near future. It is a chance to make the decision about purchasing a new product instead of maintaining one that could be outdated in a short period of time.

Other devices on the network should also be maintained appropriately. Most modern network infrastructure can be maintained using management tools that monitor them and allow an administrator to push updates to them. Some common systems are Solarwinds and Auvik. Manufacturers have their own specific tools for their products also. If you have a home or small office with just a few network devices, you can have your IT professional maintain those without any management tools. Anything beyond a few network devices requires increased vigilance and monitoring to make sure configurations are backed up regularly, devices are patched and upgraded, and no other irregularities are happening on the network. These devices should be treated just like endpoints, servers, and any other computer on the network.

Server Protection

Your company's servers are the core of your technology infrastructure. Unless you have fully transitioned to the cloud (we'll discuss those implications shortly), they hold all your data and applications. They also operate many of the critical functions of your network, such as being your internal DNS server, Dynamic Host Configuration Protocol server, and for most businesses, your Active Directory server. If your server or servers go offline, you are essentially out of business. Because of the critical nature of your server infrastructure, it is essential that they are protected on multiple levels.

Physical security is one aspect of server protection. Thinking about protecting your systems from other threats, such as loss of power, fire and water damage, hardware failure, and theft, is just as important as cybersecurity protection. We'll discuss ways to provide for these in chapters 7 and 8.

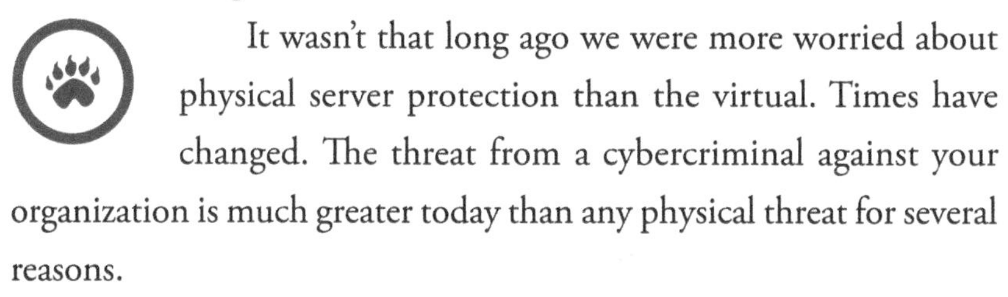

It wasn't that long ago we were more worried about physical server protection than the virtual. Times have changed. The threat from a cybercriminal against your organization is much greater today than any physical threat for several reasons.

- A hacker can take time to plan an attack once they are in your systems.
- Criminals may be after more than just taking your servers offline. They are also after your information.
- Cybercriminals use terrorist tactics against their victims, such as using ransoms to demand money.
- Virtual attacks can also ruin the reputation of your business, causing damage with your customers and the marketplace.

There are many ways for a hacker to get into your servers. We've already discussed a few of them throughout the book, but I think a good review is important. We'll break these into different sections and discuss how to protect against them. This is not an exhaustive list because hackers are always coming up with novel ways to penetrate a network.

OPEN PORTS AND DIRECT ACCESS

Servers get their names because they "serve" the network and the users. Their job is to provide functionality for the users of the network to do their work. Examples of server functions include the following:

- File services
- Authentication and security
- Database systems
- Line-of-business applications
- Remote access
- Web services

For these servers to provide their functionality to users outside the internal network, a port needs to be opened on the network's firewall to allow access to the server. We often call this a pinhole through the security of the network because it allows access from potentially anywhere directly to the company's infrastructure. This may be necessary, but it does come with some risks. If the server is not patched or maintained properly, a hacker can use this direct access to the server to exploit a vulnerability and gain access to the system. I also find that too many ports are opened for a particular service, or they are left open after a server is decommissioned, leaving literal holes in security that should have been patched.

A periodic audit of the open ports on the firewall can assist in preventing ports being opened that should not be. I strongly suggest that the absolute minimum number of ports should be opened for the entire world to access.

Another option, which is far more secure, is to require users who need access to specific systems to log in to a VPN before accessing the services. This makes sure that all data that is transmitted is encrypted over the Internet and provides for a second authentication into the systems. In this scenario, only services that need to be public, such as public websites or email services, are exposed to the public Internet. Line-of-business applications that don't require public access can be kept secure behind the firewall, and only those connected to a VPN can have access. This method also works if you are running your applications in a cloud environment.

One big issue I often find in small networks is open ports connected directly to workstations. An easy way for users to gain remote access to office computers is through Remote Desktop connections. The network administrator opens a port on the firewall to allow direct access to the computers, allowing the user to connect to the computer from anywhere. Microsoft Remote Desktop is exceptionally vulnerable to brute-force password attacks, where the hacker tries to guess a user's password repeatedly. Remote Desktop does not block login attempts after failed passwords by default. I've had to clean up too many hacked networks where the previous network administrator had done this because it was easy, allowing the network to be attacked, and the whole network received a ransomware virus.

Take the time to have your network administrator review the open ports and verify that all systems that have direct remote access are secure.

LOGGING AND SIEM

Computers track messages about what happens on the system in logs. A log stores events such as users accessing the system, programs that are running, changes made to programs or system files, and a host of other events. They also track failed login attempts, abnormal program activity, or other events that are outside the normal activity of the computer or system. On firewalls, they generally are kept for attempted intrusions, access to systems that are not authorized, or activity that is flagged by the administrator as they monitor the network.

On the surface, much of the logs that are kept are benign. A normal computer, server, or firewall contains thousands of entries in the log files that are just normal activity for the system. The logs are not very interesting until some type of event happens on the system, such as investigating a program issue, a potential hack, or other abnormal behavior.

Hackers, knowing that IT professionals are going to review the logs, generally erase them if they gain access to a computer system. It covers the tracks of the hacker so the IT professionals can't figure out how the hackers got into the system or what they did while they were in there. It also covers up where they came from so they can attempt more hacks in the future.

What if logs were kept by a central system and analyzed constantly to look for abnormalities? What if the logs from all the computers, servers, and firewalls were aggregated into one central log to look for issues that are happening across the entire network? That information would be extremely useful for a security professional to do forensic investigation on how a potential hack occurred. Even better, a monitoring system that constantly looks for abnormalities across all the logs could predict if an attack is happening in real time. That is the purpose of a security information and event management (SIEM). A SIEM collects information on all the systems across the whole network and monitors for abnormal behavior, then alerts the network administrators of the potential issues.

Remember the example of the hacker using a brute-force attack to get into a Remote Desktop connection? Those failed login attempts would be monitored by the SIEM, and a network professional would be alerted to take some type of action against the attack.

MDR AND SOC

What happens if that attack is caught by the SIEM at three o'clock on a Sunday morning and your IT professional is asleep in bed? (I'm sure your IT guy never sleeps, right?) Also, SIEMs often generate too many alerts for a normal IT team to handle. Most of the alerts are false positives, so the IT team tends to ignore most alerts, and the SIEM becomes useless.

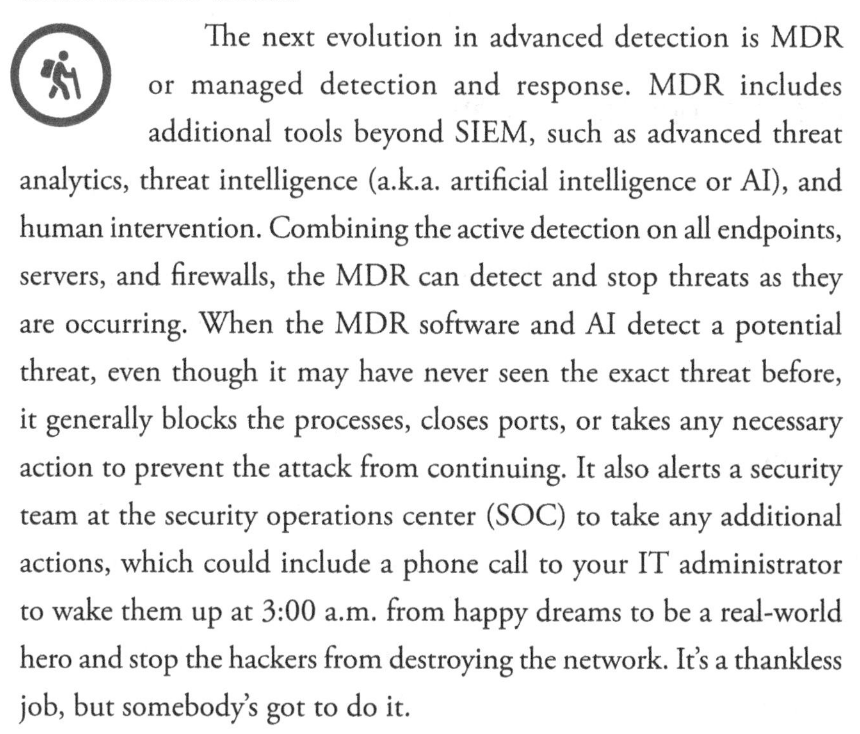

The next evolution in advanced detection is MDR or managed detection and response. MDR includes additional tools beyond SIEM, such as advanced threat analytics, threat intelligence (a.k.a. artificial intelligence or AI), and human intervention. Combining the active detection on all endpoints, servers, and firewalls, the MDR can detect and stop threats as they are occurring. When the MDR software and AI detect a potential threat, even though it may have never seen the exact threat before, it generally blocks the processes, closes ports, or takes any necessary action to prevent the attack from continuing. It also alerts a security team at the security operations center (SOC) to take any additional actions, which could include a phone call to your IT administrator to wake them up at 3:00 a.m. from happy dreams to be a real-world hero and stop the hackers from destroying the network. It's a thankless job, but somebody's got to do it.

Because the threats are increasing dramatically against businesses of all sizes, many insurance providers are now requiring these advanced layers of protection before they will write any more policies. The losses to insurance companies for cyber liability insurance are increasing exponentially, so they want companies to put strong protections in place before they will even insure them.[11] Insurance may be the last line of defense for a company, but you are going to want to get it. Following their rules will help protect your company on the first line of defense and offer you protection with insurance on the back side.

Cloud Protection

Securing cloud infrastructure is not much different from securing internal network infrastructure. There are a few considerations that should be taken, depending on the type of cloud systems that you are running: public cloud, private cloud, or hybrid cloud.

PUBLIC CLOUD PROTECTION

Since you are relying on another organization to operate the public cloud services, they are responsible for ensuring that the systems and data are secure, or they should provide the tools to allow your IT administrators to secure the systems. You should review their SLAs to ensure that they match your expectations for both uptime and security. Additionally, if your organization has compliance requirements, such as HIPAA, SOC, or PCI, the cloud environment needs to also comply with those requirements. For example, if you fall under HIPAA rules, your cloud provider should agree to and sign your business associate agreement (BAA). Also, review the agreements to clearly understand who owns the data stored in the cloud, how you can take backups of the data, and how you can transfer data off the cloud systems if you decide to move to a different provider. Finally, ensure that they have

---

11 Arielle Waldman, *Cyber insurance premiums, costs skyrocket as attacks surge*, October 11, 2021, https://www.techtarget.com/searchsecurity/news/252507932/Cyber-insurance-premiums-costs-skyrocket-as-attacks-surge (accessed March 14, 2022).

systems in place to protect against all the vulnerabilities that we've discussed in this chapter, such as DDoS protection, SIEM and MDR, and geo-redundancy of servers and services. You will pay extra for all these services, but it is well worth the money to keep your information and infrastructure secure.

It's also important to understand who has access to your servers and data. Having too many users, whether they are internal or from a vendor, with administrator credentials can lead to large security holes. Limit access to your systems to individuals who have had security training, understand the complexities of the systems, and have had background checks to ensure personal integrity. Make sure your cloud provider is taking the same precautions with staff that have access.

PRIVATE CLOUD PROTECTION

Private cloud infrastructure uses similar technologies as the public cloud, but in this case your organization has full control over all the hardware and software. The responsibility for security rests entirely with your IT administrators. Having the complete staff available to ensure the security of the systems is essential, or it is also possible to rely on trusted and highly competent outside companies to perform this service for you.

> **The responsibility for private cloud security rests entirely with your IT administrators.**

Private cloud infrastructure has the advantage of being separated from the public Internet. It is also important to keep the private cloud separate from other systems on your network. Your network team can use VLANs and firewalls to separate core infrastructure from your workstations and other systems.

That way, if a computer is compromised, it does not have the ability to take down all your servers and devices.

Periodic vulnerability assessments can assist in verifying the security of your infrastructure. Use an outside company to take the role of a red team. This group will pretend to be hackers and will work to find vulnerabilities in the systems. They will report back to your IT professionals on areas that they were able to find weaknesses in or even access the systems. Your IT team will be responsible for fixing those issues before having the group run another vulnerability assessment.

HYBRID CLOUD PROTECTION

Hybrid cloud infrastructure is a combination of private cloud and public cloud services. This technology is generally used in large environments, especially when your IT professionals want redundancy between internal systems and off-site or geo-diverse infrastructure. When using a hybrid model, your internal policies should take precedence, and your IT team should have full control over the infrastructure, adopting the same security model and tools across all the systems. Because some of the infrastructure lives in the public cloud, it is important to run those periodic vulnerability assessments to ensure that the connections between the private and public systems are secure. Additionally, there are more points of entry into the systems, so additional care is needed to secure those connections.

External Cybersecurity Checkup

Give yourself a score of 1 to 3 in each area. Add them up to see how you did overall in this area. Focus on just one area to improve upon for now.

AREA OF CONCERN	SCORE
I have reviewed and understand the SLAs by all my service providers, and they comply with my expectations of service.	
I have implemented methods to detect and prevent DoS attacks against my systems.	
All of my systems are current on the latest firmware and software revisions, including firewalls, switches, and network-connected equipment.	
My IT professionals periodically review open ports and direct access into my network and disable any unnecessary access.	
I have implemented a SIEM and an MDR. All devices are monitored by a 24/7 SOC.	
TOTAL **TOTAL POSSIBLE 15**	
NOTES	

External Cybersecurity Key Points

- There is no way to guarantee 100 percent uptime for all your systems. Discuss with your internal IT team and all your vendors what their guaranteed uptime is.
- Hackers have a number of tools to cause disruptions of your systems, especially if they have a large network of bots that can attack your network.
- DoS is a modern form of warfare. They can be used to take down companies and service providers. There are tools to prevent and mitigate against DoS attacks.
- Network equipment, such as firewalls, switches, printers, and other network-attached devices, need to be updated, just like workstations and servers.
- Allowing direct external access to your network by opening ports can lead to additional security vulnerabilities. Review the open ports periodically to ensure that only essential access is allowed.
- 24/7 monitoring of your systems by security professionals is essential in today's security environment.
- Your IT professionals are generally reactive to security issues. To be proactive, you need to have separate security professionals monitoring and testing your systems.
- Cloud environments require extra care to protect your data and systems. You need to work with your vendors to understand their responsibilities and guarantees.

CHAPTER 6

SOCIAL ENGINEERING

My first job in technology was working for a small computer store while I was in high school. I would go in for a few hours after school and help with customer service calls, shipping and receiving, and building computers. That job laid the groundwork for the company that I'd eventually start.

One afternoon I received a call from someone who sounded very professional. He told me that he worked for the company that serviced our copier but would need the model and serial number on it so he could send out the latest supplies. Being sixteen years old and overly eager to help, I promptly went over to the copier and started giving him some information. One of the other technicians heard the conversation and cut me off before I could finish. I had been the victim of social engineering and a phishing attack.

It wasn't until later in my education that I learned exactly what social engineering was. In one of my computer science classes, the professor required us to watch the movie *The Takedown*. It was a dramatized story of Kevin Mitnick, one of the first successful computer hackers who used social engineering to infiltrate large multinational companies. Mitnick once said, "I was addicted to hacking, more for the intellectual challenge, the curiosity, the seduction of adventure; not for stealing, or causing damage or writing computer viruses."[12] His intellectual curiosity led him to penetrate the networks and security of some of the largest companies in the world, mostly with little technical work. It was through his ability to use social engineering to convince people to do things that they wouldn't normally do.

"I was so successful in that line of attack that I rarely had to resort to a technical attack," Mitnick said. "Companies can spend millions of dollars toward technological protections and that's wasted if somebody can basically call someone on the telephone and either convince them to do something on the computer that lowers the computer's defenses or reveals the information they were seeking."[13]

Just like my mistake of allowing a stranger over the phone convince me to give them some information that I knew I shouldn't have given, hackers are successfully using social engineering every day to bypass all the protections that security professionals put in place to protect data and networks. Some of these attacks are very sophisticated, like a hunter that seeks and lures its prey over a long period of time, and others are just shotgun tactics that spray enough bullets to hopefully hit something. In this chapter we'll discuss the techniques of hackers and how to protect yourself and your company from them. Remember, the number one vulnerability in your company are the people who work there, so you need to learn how to stop them from unintentionally (or intentionally) destroying your business.

---

12 Cybersophia, "100+ Best Cyber Security + Hacker Quotes," accessed April 3, 2023, https://cybersophia.net/quotes/hacker-and-hacking-quotes-sayings/.

13 Associated Press, "Noted Hacker Speaks Before Senate Panel," March 2000, accessed April 3, 2023, https://www.nytimes.com/2000/03/02/technology/noted-hacker-speaks-before-senate-panel.html.

Key Definitions

- **Phishing:** Phishing is when hackers send out an untargeted attempt to solicit sensitive information from users by impersonating a legitimate entity, service, or person.
- **Spam phishing:** Sending out mass phishing attacks to a wide audience, generally nonpersonalized but can catch an unsuspecting person.
- **Spear phishing:** Using information about a person, either publicly available or purchased through the dark web, to target an individual for phishing.
- **SMS phishing:** Using text messaging to target individuals.
- **Angler phishing:** Generally occurring on social media, an attacker imitates a company to intercept communications you'd have with a support department of a known company.
- **Deepfake:** Using algorithms and machine intelligence to alter a person's voice or appearance so that it is extremely difficult to differentiate between this image/audio and the real individual.
- **Spoofing:** The process of making a call or other communication attempt seem as if it comes from another location.
- **Zero-day vulnerability:** A vulnerability in a system that is unknown by the manufacturer and has no patch or fix available. Hackers find and exploit zero-day vulnerabilities because there are limited protections against this type of attack.
- **Scraping:** Gathering data from publicly available websites, social media sites, or other locations where data is available to everyone.

Implications of an Open Society

I believe, like most Americans, that we value our freedoms and openness to the extent that we will fight for them. The ability to travel freely, choose where we live, make our own decisions, and live life the way we see fit is at the core of the nation. The United States was founded on Judeo-Christian beliefs, which include "Love thy neighbor as thyself"[14] and "Love your enemies and pray for those who persecute you."[15] These beliefs require Americans to build trust with one another.

Timothy Levine, a researcher and professor at the University of Alabama at Birmingham, studied this phenomenon in humans and developed a theory that he calls "truth-default theory." He states, "When humans communicate with other humans, we tend to operate on a default presumption that what the other person says is basically honest."[16] For the most part, humans are right in accepting this principle because we generally are honest in our communication. He goes on to say, "The presumption of honesty makes humans vulnerable to occasional deceit." The reality is that we are generally bad at telling the difference between a lie and a truth.

Have you ever played the *get-to-know-you game*, two truths and a lie? Each person in the room tells everyone else two true things about themselves and one lie. The group—people that know nothing about the other individuals—must discern which thing about them isn't true. I always find this a difficult game because usually the most outrageous thing that people say is a "truth," and small innocuous

14 (*King James Version*, Mark 12:31)

15 (*King James Version*, Matthew 5:44)

16 Timothy R. Levine, *Deception and Deception Detection*, n.d. http://timothy-levine.squarespace.com/deception# (accessed March 27, 2022).

things are generally the "lies." It is also hard to discern the difference between a truth and a lie with people that you just recently met and, more importantly, want to get to know and like.

There is the problem: we generally want to believe people that we like and want to trust. We default to truth when the person we are interacting with is likable. Kevin Mitnick states, "Successful social engineers have strong people skills. They're charming, polite, and easy to like—social traits needed for establishing *rapid rapport and trust.*"[17] The worst bad guys are not easy to spot. They generally work hard to seem like "us" and build a relationship with the person they are attacking. We let our guard down and let them into our worlds, easily revealing information that we wouldn't normally under untrusting circumstances.

There are other techniques that attackers use to gain information that we'll discuss in this chapter. Building trust is just one tool in a large war chest that they use to gather information to exploit you and your organization. My goal isn't to make you paranoid of every person on the planet but to give you a perspective of what to look out for and how you can train the people in your organization to respond to these challenges.

The Danger of "Innocuous" Information

Why did the person on the other end of my copier phone call want information about the copier for a small company? There might have been several reasons.

17 Kevin D. Mitnik, *The Art of Deception,* (Indianapolis, Indiana: Wiley Publishing, 2002).

- He just wanted to sell us some toner and supplies.
- He wanted to upsell us on a new copier if the one we had was out of warranty or old.
- He wanted to know if we had a device of value so he could break into the business and steal it.
- He wanted to know if we had a device that had vulnerabilities that he could use to get into our network.
- He wanted to build some rapport with me to get more information later.

I'm not sure what was the ultimate reason behind the phone call, but I'm sure none of these would have had the best outcome for me and my employer.

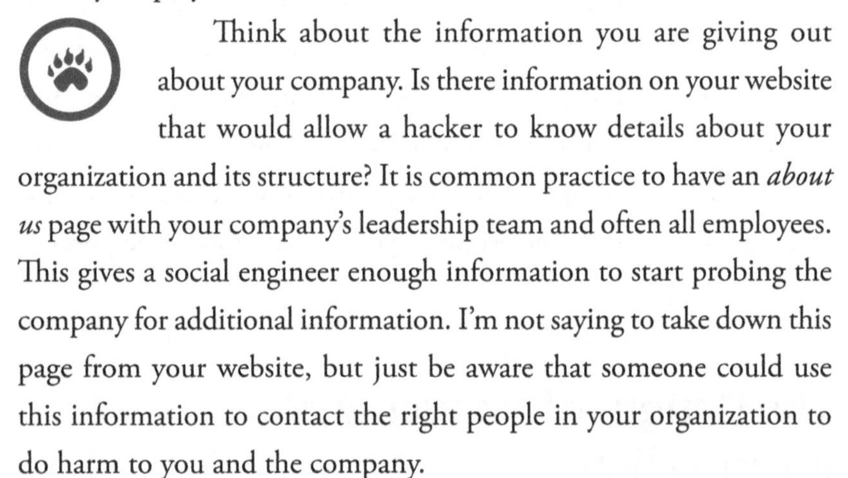

Think about the information you are giving out about your company. Is there information on your website that would allow a hacker to know details about your organization and its structure? It is common practice to have an *about us* page with your company's leadership team and often all employees. This gives a social engineer enough information to start probing the company for additional information. I'm not saying to take down this page from your website, but just be aware that someone could use this information to contact the right people in your organization to do harm to you and the company.

BE FRIENDLY, BUT ONLY TO VERIFIED FRIENDS

One of the other tricks that social engineers use to gather information about the inner workings of the company is to sound authoritative when calling or contacting the company. They will gather information from your website, including vendors that you advertise that you sell to or do business with, or they will call and ask your employees seemingly innocuous questions. The natural tendency of the people who work for you is to be helpful to callers who sound like they need help.

In his book *Way of the Wolf*, Jordan Belfort describes how he was always able get past gatekeepers to get to decision makers. Using a friendly voice, he would ask for his target by first name while sounding very sure of himself. A friend doesn't call another friend and use a full name or sound nervous. Remember that social engineers use their likability to get into organizations. They are going to call, email, or talk with surety and ask for things that the person on the other side of the conversation is going to feel confident in giving to them.

To prevent this type of attack on the organization, train your people to politely decline any requests for information until the authenticity of the request can be verified.

To prevent this type of attack on the organization, train your people to politely decline any requests for information until the authenticity of the request can be verified. This is much easier in smaller organizations where the people are working closely together, and requests can be verified by yelling down the hall to ask the person they are trying to connect with. "Hey, Suzy! Do you know Bob Simmons? He's calling again asking about that bill you're supposed to wire. Nope? That's what I thought. I'll get rid of him."

In all organizations, large and small, you should have specific procedures on disclosure of nonpublic information, even if that information seems innocuous to employees. Email addresses; direct dial phone numbers; internal procedures, especially dealing with financial or proprietary information; and any other confidential information should be guarded very closely. Also, remember that it isn't just the people who are answering your phone that will get these types of questions. Attackers will target different individuals inside your organization to paint a full picture of internal processes and information. All your employees should be trained to watch out for these types of requests and block them.

SCRAPING INFORMATION

It doesn't take much work to guess information about a company. The attacker (or better yet, the attacker's automated software) can start to guess email addresses based off information that it finds on the company's website. Let's say that your name is Alice Smith, your CFO's name is Fred Johnson, your COO's name is Jane Williams, and your marketing manager's name is Larry Miller. How does the hacker know this information? Again, you've probably told him by publishing it in a blog post, your LinkedIn profile, or some other common public information place. The hacker's software can gather all that information automatically by guessing commonly used email address formats, such as (first initial).(last name)@yourcompany.com or (first name).(last name)@yourcompany.com. The software then verifies the email address by sending the start of an email request to the mail server. That conversation goes something like this:

Attempt #1:

Scraper: Hello yourcompany.com mail server.
Email server: Hello mail sender.
Scraper: I'd like to send an email to f.johnson@yourcompany.com.
Email server: That's an unknown email address.
Scraper: Oh, sorry, never mind.

Attempt #2:

Scraper: Hello yourcompany.com mail server.

Email server: Hello mail sender.

Scraper: I'd like to send an email to fred.johnson@yourcompany.com.

Email server: Sure, what would you like to send?

Scraper: Oh, sorry, never mind.

Now the attacker knows that Fred Johnson's email address is most likely fred.johnson@yourcompany.com. The attacker can then do the same for all the other people he knows about inside the organization and gather their email addresses. Does this sound far-fetched? You might want to tell your marketing manager about it because there are many companies that offer tools like this for the marketing department to use to create marketing lists. Software such as Skrapp.io, Hunter.io, Octoparse, Prospect.io, and ScrapeBox are designed to gather information from known sources. They use databases such as LinkedIn, Facebook, Instagram, your company's website, and other public data sources to compile information on your company's organizational chart, email addresses, company size, and anything else they can use to allow your marketing department to target the right people inside your organization. Hackers can use this same information and similar tools from the dark web to find out who they should target inside an organization.

Engineering Methods

A hacker has many tools in his arsenal to attack an organization. These tools range from simple to very complex. A hacker can combine sets of tools and methods together to create different attack vectors against an individual or an entire organization. Sometimes the hacker uses a splatter method, where he attacks many individuals as often as possible to see if he can hit a target, or uses a very targeted, sniper-like approach to go after one person. Both can cause significant damage, if people inside the organization are willing or naive enough to play along.

DOMAIN SPOOFING

Before we discuss spoofing, let's talk about what a domain name is and how to get one. A domain name is an Internet name that is registered with a company on the Internet so that other people can find you. A domain name generally consists of a chosen name followed by a dot (".") and then followed by a top-level domain (TLD). Examples of TLDs are "com," "org," "info," "net," or "us," as most of you know. For example, my company's domain name is StimulusTech.com, where StimulusTech is my chosen name, and "com" is the TLD.

Getting a domain name is relatively cheap and easy. There are many companies that provide registration services for domain names. These companies are authorized by the Internet Corporation for Assigned Names and Numbers, a nonprofit organization responsible for managing the names and IP address assignments across the Internet. For about $10, anyone can register a domain name at any time. Can you quickly spot the difference between these domains?

STIMULUSTECH.COM
ST1MULUSTECH.COM
STIMU1USTECH.COM
STMULUSTECH.COM

The first one is my company's valid domain name; the other three are potential domains that a hacker could register and use to pretend to be from my company.

We used to say that the user should verify the domain name is valid when going to a website or receiving an email address. The human mind can't always see the difference between a fake and a valid domain name. Remember truth-default theory? Your mind is going to automatically assume that the domain is valid if it looks close enough. Our minds are bombarded with too much information all at once to be able to check every domain name to make sure it is valid.

Security companies have created artificial intelligence systems to detect and learn how hackers are spoofing domains. For example, email systems can alert a user about the suspected domain spoofing and alert the receiver of a potential problem. For example, this is the alert that our clients get when they are using our services:

> This email was sent from outside your organization, yet is displaying the name of someone from this organization. This often happens in phishing attempts. Please only interact with this email if you know its source and that the content is safe.

This type of alert tells the user's mind to slow down and double-check to make sure the person they are communicating with is really who they think it is. DNS filtering software, which we talked about in chapter 4, will do the same thing when browsing the Internet. Because this is a very easy way for attackers to gain the trust of the people they are targeting, software and systems are now required to catch it because users are no longer able to tell the differences between valid and invalid domain names.

SENSE OF URGENCY

I help my parents with their computer issues from time to time. One day I got a frantic text message from my mom about her computer. She told me that she had someone from Microsoft on the other line, and they were telling her that Microsoft had detected that her computer was hacked, and they needed to immediately get access to the computer to help resolve the problem. If she didn't give them access, the hackers would gain access to their bank accounts and take all their money. She didn't know what to do because she didn't want to lose any money and was about to allow the man on the other side of the phone to access her computer through some remote control software. Luckily, my mom was smart enough to message me first before doing anything. I told her to hang up the phone and that I'd be over later to check her computer (and take a peppermint patty from their candy jar for payment).

A good social engineer who is attacking your computer will use our internal fight-or-flight response by creating a strong sense of urgency. They tell their victim that if they don't do something *right at that moment*, then great harm will come to them. The con has to be believable enough and create a strong enough sense of urgency that the person will act.

Attackers will use the following:

- potential loss of money
- utilities being shut off
- pretending to do a tax audit
- vendors cutting off services

These are just a few. There are many other negative outcomes that attackers will use to convince a potential victim that they need to act quickly to prevent something terrible from occurring.

The antidote for this social engineering method is to slow down and verify the information with a reputable source. Most companies are not going to call you right before they shut off services. You usually get a certified letter in the mail or some type of posted notice well in advance of any negative action. Take a deep breath and contact the source directly to verify the claim. Call your utility company with the number that is found on the monthly bill. Use the phone number on the back of your credit card. Contact your vendor with the number you have stored in your accounting system. *Never* use the number or email that the attacker is using to contact you. It is most likely *spoofed*, and you will just be contacting the attacker if you do.

BAITING

Did curiosity kill the cat? I think curiosity killed the cat's owner because humans are naturally very curious, and hackers know it. Do you want to have some fun and do a bit of research? Leave a USB thumb drive in a public area and see if anyone is willing to pick it up and plug it into their computer. Maybe people want to be helpful and return a potentially lost drive. Or maybe they are just curious to see what is on the drive. Either way, plugging a random USB drive into a computer can bring down a country's nuclear enrichment program or a government satellite.[18]

Wait, *what*? This story went from zero to sixty in about two seconds! I did just say that a lost USB drive has the potential of causing

---

18 Bruce Schneier, *The Story Behind The Stuxnet Virus*, October 7, 2010, https://www.forbes.com/2010/10/06/iran-nuclear-computer-technology-security-stuxnet-worm.html (accessed April 19, 2022).

a major infection that could bring down government systems. Once the drive is plugged into a Windows computer, the computer will automatically run a program on the drive and infect the computer. At the time the Stuxnet virus exploited a zero-day vulnerability in Windows that allowed it to infect the computer and then scan the network for industrial systems to infect. It didn't do any harm to the computer it was plugged into but caused major issues to the networks and equipment it did infect. It's estimated that fifty-five thousand computers worldwide were infected with the Stuxnet virus.

Baiting takes many forms other than just a random USB thumb drive. It could be an advertisement for a special deal or coupon. It could be a chance to win money or go on a dream vacation. Baiting will always focus on our curious or helpful nature. Often, a bait is either *too good to pass up* or *too good to be true*.

Don't fall victim to the sense of urgency from your caller. The best defense against baiting is to *pause and think*. And if you find a USB drive lying around, give it to your computer professional (and pray that he doesn't just plug it in). Call the company that is offering the special deal. Use second methods to verify that the information that you are presented is valid.

PRETEXTING

Using our natural desire to be helpful, an attacker is going to use their friendliness to craft a set of lies to get their victims to disclose information that will be useful in a later attack. This set of lies is called *pretexting*. I mentioned earlier how an attacker gathers "innocuous" information about a person or organization. In this case the social engineer is using pretexting to gather the information. It could be by impersonating someone with authority, such as a police officer, an FBI agent, or a computer expert. It also could be by pretending to be a vendor, someone conducting a survey, or a new coworker. A sophisticated attacker uses pretexting to gather as much information about the target so that they can find the right targets in an organization, act like an insider, and gain access to sensitive information such as bank account information, social security numbers, or passwords.

The best way to protect against a hacker's ability to gather this information about you and your organization is to be *very suspicious* and verify their identity through direct sources.

How to do this:

- Provide a quick awareness training session for everyone who works at your organization.
- Let everyone know they must never take the attacker's word about their credentials.
- Advise every employee to never disclose private or confidential information.

Phishing

Once an attacker has information about you, your employees, and your organizational structure, there are several ways he can proceed to engineer himself inside your organization. The simplest way to get inside an organization is to get the employees to let the attacker in.

I'll give you an example of how quickly hackers can get access to this information. My information and my company's organizational structure is on the dark web. Over the years several prominent websites have been hacked, such as LinkedIn and Dropbox. Once that information is out there, it is impossible to get it back. All of us just have to assume that at least our names, email addresses, and phone numbers are accessible by anyone who wants that information. As an example, my company hired a new employee in another state. About one week after starting with the company, the new team member changed his LinkedIn profile to show that he was employed by Stimulus Technologies. This is a logical and common thing to do, and I'm proud that he is excited to be associated with the company. Within hours, possibly sooner, he received a text message from me asking him to do me a favor. Unfortunately, it wasn't me, and it wasn't my phone number. Being helpful, he said that he was happy to help. He and I had spoken through the interview process, and I would have had access to his cell phone number through our human resources system if I wanted to contact him directly. The attacker asked him then to help me out since I was traveling (which was also true because I had an out-of-office message on my email). When the attacker asked him to go get some gift cards so I could take them to a few clients, he immediately became suspicious and stopped communicating to the attacker. He then reported it to his supervisor and, in turn, to our internal security team. The only thing that surprised us was the speed with which the attacker found his information change on LinkedIn and contacted him. There is little that technology could do to prevent this, other than the new team member changing his LinkedIn profile to private so it could not be scraped again. He also recognized that his cellular phone number is out there on the dark web and associated with his name. Once the Pandora's box is open, it becomes nearly impossible to shut.

PHISHING WITH A BIG NET

The example of the "new employee" is a big net phishing attempt. The attacker is casting as wide of a net as possible inside my organization to see if anyone gets caught. This becomes a numbers game, which is one that hackers win every day. The more attacks they make, the higher the probability they have to catch someone. It is like gambling, eventually the hacker will hit a jackpot, and they have unlimited funds to play with. The hackers may use a number of social engineering methods to achieve their goals, such as creating a sense of urgency, like your power will be shut off, an invalid charge on a credit card that you have to verify, or a quick favor to help a friend or your boss. These usually come in through email because it is cheap and easy. They could also come from invalid search engine results (you mistype a name of a company and you get the hacker's website), phone calls, or text messages. The hackers will use any number of ways to contact as many people as possible until they get someone to take the bait.

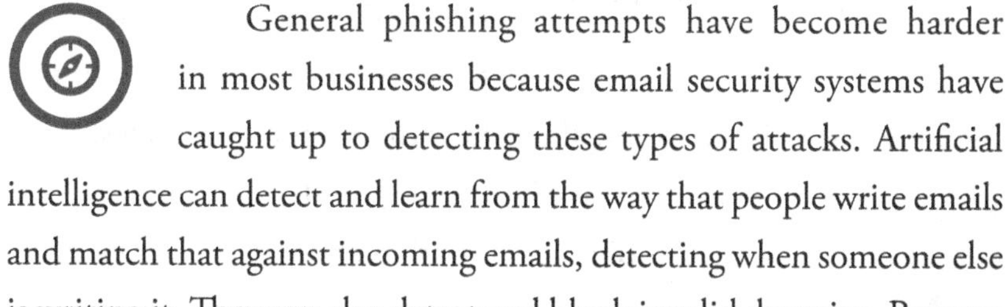

General phishing attempts have become harder in most businesses because email security systems have caught up to detecting these types of attacks. Artificial intelligence can detect and learn from the way that people write emails and match that against incoming emails, detecting when someone else is writing it. They can also detect and block invalid domains. Because these systems are cloud-based, they also can work together to detect high quantities of the same email or types of emails from one source. Companies should ensure that their email system has advanced protections in place because it has become too difficult for people to detect threats because of the quantity and quality of them. Humans should only be the last line of defense against phishing, not the first.

SPEAR PHISHING

Whereas phishing with a big net is a numbers game, spear phishing is far more precise. Attackers know that companies and individuals have learned and implemented many systems to prevent people from falling for general phishing attempts, and the "Prince from Nigeria" can't do business in the United States anymore. (This is an old email phishing scam that has been around for as long as I can remember.)

When an attacker wants to attack a specific organization or person, they generally revert to more sophisticated tactics, which takes more time but can have a greater payoff. Like I mentioned above, hackers can learn all about any organization by scraping information about them from their digital assets. More importantly, they can get information by using baiting and pretexting to carefully gather information about how the company operates.

When an attacker wants to attack a specific organization or person, they generally revert to more sophisticated tactics, which takes more time but can have a greater payoff.

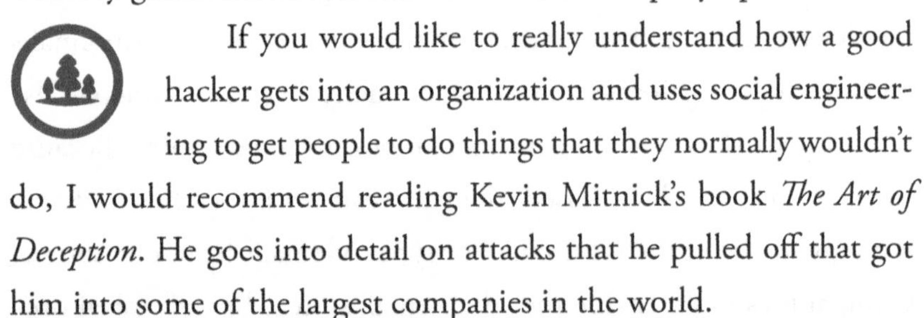

If you would like to really understand how a good hacker gets into an organization and uses social engineering to get people to do things that they normally wouldn't do, I would recommend reading Kevin Mitnick's book *The Art of Deception*. He goes into detail on attacks that he pulled off that got him into some of the largest companies in the world.

Let's say that an attacker wants to get a company to wire them some money. A spear-phishing attack may go something like this. The attacker comes up with a plan to get the company's controller to wire funds to a vendor to pay a large invoice. The attacker would need to collect information about the vendors the company normally makes payments to, how they pay those bills, possibly which bank they use, and how they approve wire transfers. This type of attack is very sophisticated because the attacker will need to take the time to learn, through pretexting, all this information. They may need to gather it by pretending to be a vendor, so they will try to find out how to send invoices to the company. They would need to possibly hack any random employee's email box to get access to pay stubs to find the bank's routing numbers. They would also need to monitor the owner's out-of-office messages, social media posts, or other information about her travel details. Once all the details are in place, they could launch an attack on the controller, pretending to be the owner and requesting a vendor to be paid. Unless the company has strict verification policies in place, these types of attacks happen often and have huge consequences. I have seen many good companies, even with the right controls in place, be taken by a sophisticated attack. Strict diligence to the right controls and policies are the only way to fully protect against a strong spear-phishing attack.

OTHER TYPES OF PHISHING

As I mentioned before, hackers are going to use more types of media to attack organizations. Be on the lookout for phishing from text messages, phone calls, or even post office mail. Be aware that hackers are using as many tools as they can to get into your organization.

Last Lines of Defense

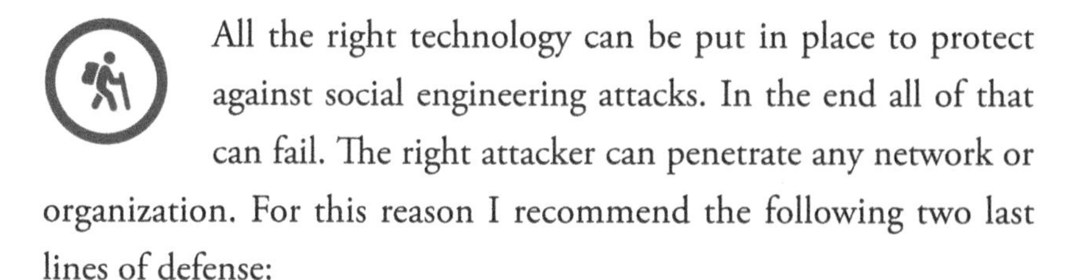

All the right technology can be put in place to protect against social engineering attacks. In the end all of that can fail. The right attacker can penetrate any network or organization. For this reason I recommend the following two last lines of defense:

- Continuous ongoing education for your employees. Annual security training, weekly micro training, and testing of the training will help keep your employees alert to and aware of the dangers. With persistence and education, you can break away from the "default to truth" tendency and make your people a little more paranoid of the potential threats.

- The right insurance for your company. Have a clear discussion with your insurance agent about how well cybersecurity and employee theft policies will protect your organization. Make sure the policy limits are sufficient to cover your losses *when* your organization falls victim to an attack.

The recommendations in this chapter will change over time because the tools and methods attackers use will change over time. Visit https://nathanwhittacre.com/resources to get the latest information on how to protect yourself and your organization from these threats.

Social Engineering Checkup

Give yourself a score of 1 to 3 in each area. Add them up to see how you did overall in this area. Focus on just one area to improve upon for now.

AREA OF CONCERN	SCORE
My company doesn't release innocuous information without secondary verification of the person requesting it.	
Advanced phishing detection systems are in place on the company's email systems and on all web browsers.	
I meet at least once a year with my insurance agent to make sure my company has the right cybersecurity and employee theft policies.	
I've implemented policies and controls to prevent employees from making financial decisions on their own without secondary authorization.	
My organization trains employees on detecting and preventing social engineering attacks.	
TOTAL **TOTAL POSSIBLE 15**	
NOTES	

Social Engineering Key Points

- Humans have a "default to truth" and lean toward believing an individual is telling the truth.
- Hackers who use social engineering techniques are friendly people.
- Anyone can fall victim to social engineering.
- Attackers have many tools to be able to get someone to do something they wouldn't necessarily do, including creating a sense of urgency, spoofing, baiting, and pretexting.
- Hackers can use the law of large numbers to cast a wide net to get into an organization.
- An attacker will spend months and years creating sophisticated attacks to get into their target organization.
- You can protect yourself with the last lines of defense of education and insurance.

CHAPTER 7

PHYSICAL SECURITY

My introduction to the need for good *physical* security came early on in my career. One of our clients informed us that a few disgruntled employees had left the company. They were worried that these employees might try to do something to harm the business. Our client had a reasonable level of security for their offices such as physical alarms, locks on the outside doors and separate locks on the server room, and some basic surveillance. The employees who left were managers, so they were familiar with all the measures the company had put in place. We discussed some options and quickly put a plan in place to allow them to continue operations in the event that someone broke into their offices.

A few weeks went by, and I received a call from my client that someone had broken into their offices. The intruders went straight to the server room and stole their main server. They took that and not much else because they knew they only had a few minutes to get in, take the most damaging thing for the business, and leave before the police arrived. The call wasn't one of panic but of gratitude. The plan we put in place worked perfectly, and they were able to continue operations without any issues.

What was our plan? We had decided to purchase a redundant server and place the machine in another office. The server was built to look like a workstation, and no one knew that it was a server. We designed the system to replicate all the data from the primary server every hour and detect if the main server went down. In that event it would automatically switch to be the primary server, and they could continue operations with limited loss of data, if any. We did all this before there were options for cloud storage or any of the modern techniques used today to help businesses survive these kinds of disasters.

In the era of hackers doing more damage through digital attacks, physical security is often overlooked. The things we are going to discuss in this chapter are just as important today as they were in the 1990s, when I started running my technology business. A power outage, an equipment theft, fire, or a storm damage can still affect a business and have disastrous results.

Key Definitions

- **Computer power supply:** A part of the server or computer that converts AC power supplied by most buildings to DC power that can be used by the server or computer.
- **Uninterruptible power supply (UPS):** A battery system that supplies power to equipment if outside power is cut off.
- **Surge suppressor:** A device that protects equipment from surges of high electrical voltage.
- **Volt-amps (VA):** A measurement of the power requirement of equipment calculated by multiplying the voltage by the amperage.
- **Hard drives:** Storage devices inside computers or servers that use a mechanical platter to store data magnetically and an arm that reads the data. It operates like a record player but turns at high speeds.
- **Solid-state drives:** Storage devices inside computers and servers that use integrated circuits and flash memory to store data. They are usually much faster than hard drives at reading data but have a limited lifetime on the number of writes.
- **Redundant array of independent disks (RAID):** A set of drives (hard drives or SSDs) that operate together to provide redundancy if one or more of the drives has errors or fails to operate normally.
- **Remote monitoring and management (RMM):** A tool that allows IT professionals to monitor any number of systems and receive alerts, issues, or errors in one single system. It also allows the administrators to send updates, enforce policies, and script changes on all the systems at once.

Server Management

Although some companies have moved their infrastructure to the cloud and no longer have a server on the premises, many still have critical servers and networking equipment in their offices that their employees rely on each day to do their work. It is important that companies protect their servers and network equipment from the physical dangers that exist, not just the virtual ones. There are several key strategies that companies can implement to obtain a high level of uptime of their equipment. You also may want to review chapter 2, where we discussed setting a hardware life cycle. Replacing server and networking equipment according to a schedule is part of the physical protection of the equipment.

SERVER HARDWARE DESIGN

Working with your IT team and choosing the rightsized server for your company is very important. Considerations for choosing the brand, model, and capacity are out of the scope of this book. They change consistently. Work with your professionals to pick the right equipment.

Some considerations that should be incorporated into all servers, regardless of performance requirements, include redundant power supplies, redundant drives, and business-grade equipment.

All servers should have redundant power supplies. Power supplies generate a significant amount of heat during the conversion process from AC to DC power and contain hardware that can be burned out because of surges, overpower, or overheating. Having at least two power supplies in each server reduces the likelihood of a server outage due to power supply failure. Additionally, each power supply should be connected to a separate UPS (discussed later) so that if a single UPS goes offline for any reason, the server can stay online.

Another point of failure for servers is their storage. Whether you use hard drives or SSDs, they all have a rated MTBF (see chapter 2). That means the manufacturer knows there is a definitive time that the device is going to fail but that it could happen sooner. Running the drives in a RAID, or redundant array of independent disks, allows for one or more drives to fail without taking the server down. The following are the different configurations of RAID drives:

- RAID 0: Never use this in servers. This setting does not provide redundancy; it just increases capacity by combining the drives to appear as one unit. If one of the drives fails, all the data in the entire array is potentially lost.
- RAID 1: Also known as mirroring. This is the most basic RAID setup where data is written exactly to both drives. This can only have exactly two drives in the RAID, and if one fails, the second one will continue operating normally until the failed drive can be replaced.
- RAID 10: This is a combination of RAID 0 and RAID 1. This setup must have an even number of drives, where each pair of drives is mirrored, and then capacity is increased by combining sets of drives. For example, if you have eight drives with 1 terabyte (TB) of capacity each in the array, you will have a single RAID array of 4 TB storage available. In other words you lose half of the storage capacity to the mirroring. Multiple drives can fail in the array (up to a maximum of half the drives), but two drive failures could also cause the storage to go offline. (Ask your IT professional why, and they'll be sure to give you the complicated explanation.)

- RAID 5: This type of array uses a portion of each drive in the array for redundancy. This RAID must have at least three drives in the array, and one drive's capacity is lost for redundancy. For example, if you have the 8 x 1 TB drives in RAID 5, you will have 7 TB of available storage. Also, note that RAID 5 is slower on performance than RAID 10. You could only have one drive failure before the unit goes offline.
- RAID 6: This is similar to RAID 5 but has a second drive for redundancy. Up to two drives (any of them) can fail, and the unit will stay online.
- Other RAID types: There are several other RAID types developed by hardware and software companies. Each has their benefits and trade-offs. Discuss the implications of these options with your IT professionals.

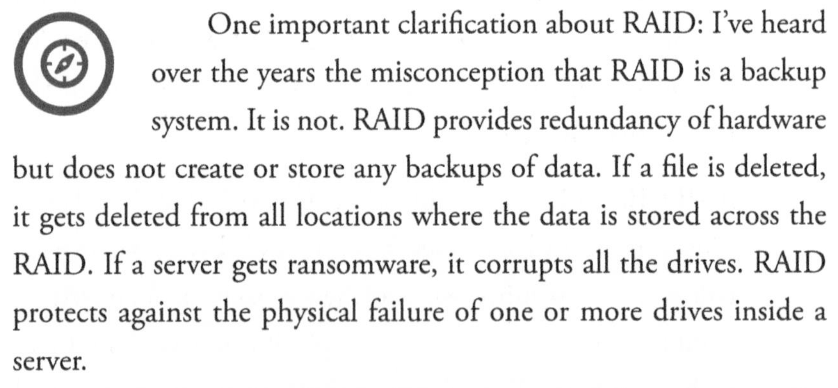

One important clarification about RAID: I've heard over the years the misconception that RAID is a backup system. It is not. RAID provides redundancy of hardware but does not create or store any backups of data. If a file is deleted, it gets deleted from all locations where the data is stored across the RAID. If a server gets ransomware, it corrupts all the drives. RAID protects against the physical failure of one or more drives inside a server.

A final note about RAID: Your IT team needs to monitor the RAID to ensure that drives are replaced when errors or failures occur. Leaving a failed drive in an array can lead to corruption of data and the eventual full failure of the array. Make sure you discuss the implications of the RAID setup in your server with your IT team and ensure that they are monitoring it.

WARRANTIES AND SPARES

Even with all the protections in place, servers die and sometimes suddenly and unexpectedly. As they are an expensive capital investment, placing and maintaining a warranty on them is important. Most manufacturers will offer next-day on-site warranty repair. We recommend keeping this active for the life of the server. It ensures that the manufacturer is maintaining replacement parts for the server for the duration of the life of the equipment. Usually, when the manufacturer no longer keeps replacement parts for a server in their warehouses, they will no longer sell additional warranties. The maximum warranty period is usually between five and seven years. It is money well spent.

We'll discuss business continuity in the next chapter, which will cover the period of a hardware failure and repair. If your business can operate for about twenty-four to forty-eight hours without the server online, a warranty is sufficient. If your business would effectively be shut down without the server in place, you should consider having spare equipment and a business continuity plan in place.

HARDWARE MONITORING

What good does it do if your systems have all these redundancies built into them and you aren't aware of a failure? If you have two power supplies in your system and one of them goes out, and you aren't aware of the failure, then you have no redundancy. Your servers should be set up to provide alerts to your IT team to notify them of failures. Most business and enterprise class servers have built-in systems to alert your team about failures through email or other alerting mechanisms, but they need to be configured correctly. In my company, we prefer to install a separate software system that monitors the hardware. This software is called a remote monitoring and management tool (RMM). The software then reports back to one alerting system for all the issues that are happening on any system we manage. It automatically creates tickets for our technicians to resolve issues, such as replacing hard drives, fans, power supplies, or other hardware problems before they become a larger issue.

Power Management

We'd love to assume that the power coming out of the wall is always going to be on and be "clean." Unfortunately, there are often power issues associated with storms, high temperatures, high energy usage, vehicular accidents, or other incidents that can cause catastrophic power issues. Generally, "dirty" power—which I classify as low, high, or varying voltage; current spikes; nonstandard AC sine waves; or short bursts of loss of power—will lead to equipment failure, lower life spans, and potential outages. Although out of the scope of this book, it is always good to have an electrician test your power before installing critical equipment.

A few years ago, a large thunderstorm came through Las Vegas, where I live, and was causing sporadic power outages throughout the city. Our servers were on battery backup, but the power company was reporting that the outage might be longer. I went into the office to hook up a portable generator to run the systems temporarily until the power could be restored. As I was working on getting the generator set up, I saw the power company's vehicles come by the office to start working on the power lines behind my building. Shortly after they arrived, I saw a lightning bolt strike one of the large utility poles, and the transformer mounted on the pole blew up. Luckily, no one was injured, but the damage was done, and I knew I'd be without power for more than a few minutes. The generators ran throughout the night and kept the power on for our critical equipment. The next day I began the process of ordering a permanent power generator with automatic transfer switches to supply power indefinitely in the event of a power loss.

For most businesses, having battery backup systems, also known as UPSs, is sufficient for their operations. The battery will keep the server and networking equipment online in the event of a power outage or even short periods of power fluctuations. Most advanced systems will run all power through the batteries, a process that cleans up the power. Having a UPS that provides pure sine wave power to equipment may also be required on some systems that have high-efficiency power supplies.[19]

Here are some considerations in purchasing and maintaining UPS systems:

- The average life expectancy of the batteries in a UPS is approximately three years. The batteries should be tested periodically using the internal testing mechanism of the UPS and replaced on a standard schedule, which I would recommend being at most three years.

- There are several types of UPS, including line interactive, online, and pure sine wave. There are some advantages and disadvantages to each type.

    - Line interactive: These generally do not run the equipment through the batteries full time, thus increasing the life span of the batteries. In the event of a power outage, it takes a short time (unnoticeable to a person)

19 Tripp Lite, *What Type of UPS Do I Need?*, January 20, 2022, https://blog.tripplite.com/what-type-of-ups-do-i-need (accessed May 12, 2022).

to switch from line power to battery power. With some equipment, especially high-efficiency server equipment, this can cause a momentary loss of power, causing the equipment to reboot. Line-interactive equipment is generally okay for networking equipment, desktops, and other noncritical equipment.

- Online: These UPS units run the power through the batteries full time, so in the event of a power outage, there is no loss in power to the equipment. The battery life for these systems is much shorter than line interactive, and the cost is generally higher. Consideration needs to be taken when replacing the batteries, as there may be specific procedures required for switching to bypass power during the procedure.
- Pure sine wave: These UPS systems ensure that the AC power supplied to the equipment is an exact sine wave. This may be required for certain highly sensitive equipment such as medical devices and high-efficiency servers. Check with the equipment manufacturer to find out if this technology is required for the installation.

• Most UPS systems are rated in both VA and watts. The difference between the rated VA output and watts output is the efficiency of the UPS. I generally size the UPS systems using watts, which is the lower rated number. To properly size the UPS system, you will need to do the following:

- Add the maximum watt requirement of all equipment that will be connected to the UPS.

- Ensure that the watt requirement is less than the maximum watt rating of the UPS.
- Verify that the runtime supplied by the UPS for the power usage meets your power budget.

UPS 1 POWER BUDGET	
Equipment	**Power Requirement (Watts)**
Server 1 (power supply 1)	470 watts
Server 2 (power supply 1)	320 watts
Router	150 watts
Switch 1	350 watts
UPS 1 TOTAL POWER REQUIRED	**1,290 WATTS**

UPS 2 POWER BUDGET	
Equipment	**Power Requirement (Watts)**
Server 1 (power supply 2)	470 watts
Server 2 (power supply 2)	320 watts
Switch 2	350 watts
DVR	300 watts
UPS 2 TOTAL POWER REQUIRED	**1,440 WATTS**

Connect the UPS to all the servers or equipment that it is supplying power to through network monitoring or USB. You will need to configure the UPS software to interface with the computer's operating system to ensure that the computer shuts down properly in the event of an extended power outage. For example, if the power has been offline for ten minutes and the unit is rated to sustain power for fifteen minutes, the UPS can send a signal to the server to start the shutdown process. This will allow the server to shut down normally and not corrupt critical data.

One final reminder about UPS systems is that the batteries do fail. You can have all this in place and still have systems go down during a power outage. Put stickers or labels on the units showing when they were installed or when batteries were last replaced so you can easily see when they need to be changed.

Environmental Design

Servers and network equipment produce heat—and a lot of it. As equipment has shrunk in size and increased in performance, the heat signatures have increased dramatically. One issue I often find is that companies put their servers and equipment in a "closet" and do not properly ventilate or cool the equipment. The closet has an air vent in it, but often, the temperature is controlled by a thermostat that is outside the room. In the wintertime the heater turns on and blows in hot air rather than cold and warms the already hot server closet.

I strongly recommend that companies install dedicated cooling units in their server equipment room. One option that I often recommend is called a mini-split system. The blower unit is in the room, and a few small pipes go to the cooling unit on the roof. This provides a dedicated cooling system to the servers that is independent from the rest of the heating and cooling systems of the building. This keeps the servers cool and the people comfortable.

One question I'm often asked is how cool to keep the server room. Historically, the industry thought that we needed to keep the equipment supercooled (between 50°F and 60°F) to increase the life span of the equipment. Recent studies[20] have shown that servers can handle much higher temperatures. The issue is that if a cooling unit failed, how long do you have to replace it before the temperatures get too high for the server to handle? To balance these two demands out, I recommend keeping the server rooms around 70°F–72°F and monitoring the temperatures using remote tools (remember the RMM). Most servers will report if their temperatures are rising abnormally and allow the administrators to take action. Another option is to install an environmental monitoring system that alerts administrators of high temperatures, too much moisture, or power outages. These units allow the team to resolve environmental issues before they become a *real* issue.

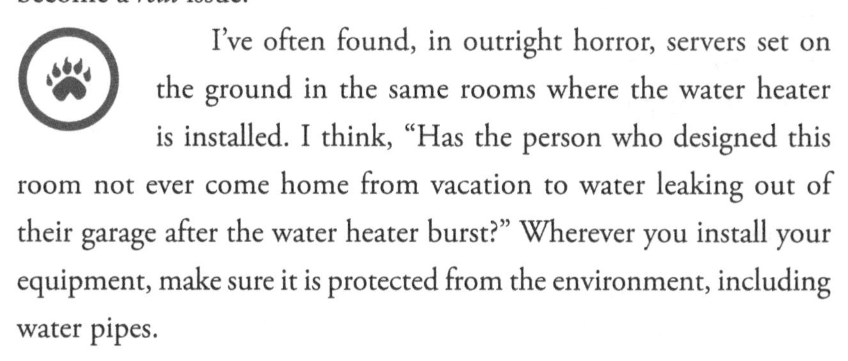

I've often found, in outright horror, servers set on the ground in the same rooms where the water heater is installed. I think, "Has the person who designed this room not ever come home from vacation to water leaking out of their garage after the water heater burst?" Wherever you install your equipment, make sure it is protected from the environment, including water pipes.

Access Control and Surveillance

Just like the story at the beginning of this chapter, physical security is still a real issue for companies. Theft, arson, and physical destruction can bring down a company's operations just as easily as cyber theft. I have often used the warning "There is no server store down the street" to tell customers that replacing stolen or damaged hardware is a real

---

20 Rich Miller, *Too Hot for Humans, But Google Servers Keep Humming,* March 23, 2012, https://www.datacenterknowledge.com/archives/2012/03/23/too-hot-for-humans-but-google-servers-keep-humming (accessed May 13, 2022).

issue. Supply chain issues that have emerged postpandemic have made lead times on electronics very long, sometimes taking months to get critical pieces of equipment.

As a minimum level of security, I recommend making sure a limited number of people have access to a server room. This can be as simple as a locked door that only owners and limited managers have access to. The door should always remain locked and only be opened to authorized people. Additional options include access control systems that log access through keypads. As with everything that is technology related, these systems should be audited frequently to ensure that terminated employees no longer have access to the building. Additionally, it's important to remember that older systems can easily be hacked. For less than $20 online, you can purchase a handheld scanner that can read keycards and duplicate them. Newer systems have rolling codes that cannot be duplicated as easily.

As a minimum level of security, I recommend making sure a limited number of people have access to a server room.

Going back to social engineering, bad actors can surveil a company by posing to be from a service provider, such as the power company, ISP, or other company that may potentially need access. Always verify that a service provider was called out to work on equipment, and be sure to monitor the work that they are doing. Do not just grant access to critical equipment to anyone with a logoed shirt or a badge.

I've found that surveillance is more of a deterrent than having any ability to catch someone who has done harm to a company. There are many options for companies to install cameras and other surveillance equipment. Much of this equipment needs network access for recording and remote monitoring. I recommend installing this equipment on a separate VLAN so the equipment is segregated from the rest of the devices on the network. Just like all other network equipment, cameras, DVRs, and other surveillance devices need periodic updates to firmware and software. My experience is that the company that is supposed to maintain this equipment does not always keep the devices up to date. Having them on the same network as your computers and servers can lead to potential security issues.[21]

21 Cybersecurity and Infrastructure Security Agency, *ICS Advisory (ICSA-21-208-03)*, July 27, 2021, https://www.cisa.gov/uscert/ics/advisories/icsa-21-208-03 (accessed May 17, 2022).

Physical Security Checkup

Give yourself a score of 1 to 3 in each area. Add them up to see how you did overall in this area. Focus on just one area to improve upon for now.

AREA OF CONCERN	SCORE
My company has a hardware replacement life cycle for all servers and battery backup systems.	
Warranties are active on all server equipment with next-day, on-site repair or replacement options.	
I have calculated the appropriate power budget for all battery backups and understand the battery runtime in the event of a power failure.	
The company's server and equipment room have separate environmental systems that are monitored for failure.	
The company has appropriate access control systems to critical equipment.	
TOTAL **TOTAL POSSIBLE 15**	
NOTES	

Physical Security Key Points

- Physical security of equipment is just as important as cybersecurity.
- Servers and critical equipment should be purchased and maintained to high standards.
- Servers should have redundant power supplies and physical storage.
- Warranties on servers and critical network equipment should be maintained throughout the life cycle of the equipment.
- Expect power to fail periodically. Companies should have sufficient battery or generated power to keep critical equipment online during power outages.
- Servers and network equipment generate heat year-round and should have separate environmental systems from the rest of the building.
- Experts in building security and access control can help companies rightsize security systems.
- Cameras and other surveillance equipment are good deterrents from theft and physical damage.
- Surveillance equipment should be segregated from the rest of the network to prevent cyber threats from the equipment to the rest of the network.

CHAPTER 8

NETWORK DESIGN AND BUSINESS CONTINUITY

Murphy's fourth law of computer science states, "If there is a possibility of several things going wrong, the one that will cause the most damage will be the one to go wrong." And to go along with it, the corollary, "If there is a worse time for something to go wrong, it will happen then."[22]

In chapters 2 through 7, I've discussed all the threats to an organization and ways that you can protect yourself and your company. The unfortunate truth is that you can do everything right and still have a breach. In this chapter we will discuss some additional designs that you can use to provide redundancy to your network, security for wireless networks, and in the event that everything else breaks, having a business continuity plan to allow you to continue to operate.

---

22 Fabien Gandon, *Murphy's general laws.* n.d. https://www.cs.cmu.edu/~fgandon/miscellaneous/murphy/ (accessed May 19, 2022).

This topic reminds me of my flight training. In 2009 I studied for and received my private pilot's license. A large part of the initial training is preparing for emergencies. The Federal Aviation Administration also requires pilots to receive recurrent training at least once every two years. The club that I belong to requires training at least once a year for an additional level of safety for its member pilots. Aircraft, including the small aircraft I fly, have numerous redundancies. The likelihood of complete failure is very low, but it does happen. Of all the general aviation (private aircraft) accidents, only 19 percent are related to mechanical failures.[23]

A few years ago, I went flying with an instructor for my annual flight review. We flew out over a desert area south of Las Vegas and practiced several flight maneuvers. After completing these maneuvers, the instructor shut the power off to the engine and informed me that I just had an (simulated) engine failure, which he initiated by closing the throttle. This reduces the engine power to idle, and the plane acts like the engine isn't working anymore (although it is easily fixed by increasing the throttle again). I quickly went through my mental checklist to attempt to resolve the issue and proceeded to look for a suitable landing spot. After deciding on the landing spot, I went through the written checklist to verify that I completed all the required items to attempt to restart the engine, all while flying the aircraft within a few hundred feet of the ground in a practice "engine out" landing in the desert area. Luckily, I was able to restart the engine (which required me to convince the instructor to take his hand off the throttle—it's a good thing he didn't bring a bottle of glue to hold his hand there permanently) and return to a normal altitude.

---

23 AOPA, *The 31st Joseph T. Nall Report*, n.d. https://www.aopa.org/training-and-safety/air-safety-institute/accident-analysis/joseph-t-nall-report/nall-report-figure-view?category=all&year=2019&condition=all&report=true (accessed May 20, 2022).

The next day I took the same aircraft on a flight to our offices in California. Upon reaching cruise altitude, I received an alert that the fuel pressure was low. This is a common occurrence at high altitudes in this Diamond DA40 aircraft, and the operating handbook specifies to engage the secondary fuel pump, which I proceeded to do. About thirty seconds after engaging that pump, the alert came on again, and the engine started to shudder and lose power. I verified that all the controls were normal and then immediately did the same procedures that I previously practiced, turning the aircraft toward the dry lake bed that I had intended to land on during the simulated emergency the day before. I communicated with air traffic control about my emergency, and they dispatched the emergency response vehicles to my intended landing site. During the whole ordeal, I was able to stay calm, executed what I was trained on, and landed the aircraft without any damage to it or myself.

After landing the plane, I found out that the secondary pump had failed, sending small metal pieces into the fuel system, which blocked proper fuel flow to the engine. Both primary and secondary systems failed, and I was left with an emergency. Because of proper training in emergency procedures, my life was not at risk, and I had a positive outcome.

You can have the same experience of avoiding technological disaster. It's possible to achieve a positive outcome with your backup systems.

You can have the same experience of avoiding technological disaster. It's possible to achieve a positive outcome with your backup systems. This chapter will talk about redundancies you can install in your network design and the types of systems you can have in place to provide you with what we call business continuity. As the name implies, with these systems in place, you can still be safe and protected even if Murphy comes to visit and everything else goes completely wrong.

Key Definitions

- **Symmetric connection:** The upload and download (or sending and receiving speeds of the port) speeds of the network connection are equal.
- **Asymmetric connection:** The upload and download speeds of the network connection are different.
- **Buffer or queue:** A memory location where data is held between network connections when more data is expected to be sent than the speed available.
- **Latency:** The amount of time it takes to transmit data between one location and another.
- **Jitter:** A measure of the variation of latency.
- **Throughput:** The amount of data that can be sent over a given connection.
- **Quality of service (QoS):** A measure of the quality of the network connection.
- **Software-defined wide area network (SD-WAN):** When multiple Internet connections are linked together into a wider network via software.
- **Local area network (LAN):** A private network where computers can communicate with each other.
- **Router:** A device that allows for communication between two different networks, usually between the Internet and a private network.
- **Firewall:** A layer of security that determines which inward and outbound Internet connections are allowed in a LAN.

- **Virtual local area network (VLAN):** Allows a network engineer to segment a network to protect critical infrastructure from other devices.
- **Recovery point objective (RPO):** The maximum length of time between a disruption of service and the last backup image (i.e., the amount of acceptable lost data).
- **Recovery time objective (RTO):** The maximum amount of time that service can be offline between a disruption of service and restoration.
- **Business continuity:** Allowing a company to continue operating normally even if primary systems go offline.
- **Failover:** The ability of a system to automatically migrate from one service to another with little or no disruption.
- **Data backup:** Copying data from a primary system to a secondary location for its preservation in the event of a disruption of service.
- **Snapshot:** Taking an image of data or a system at a single point in time.
- **Archive:** Offline data storage consisting of snapshots of data to be held for a specific period of time.

Types of Internet Services

An Internet connection is basically a bridge between two networks that allows them to communicate with each other under specific rules. The computers and services on your company's local area network (LAN) want to connect to other servers on other networks throughout the world, whether that is to send emails, communicate through messaging systems, purchase products from a company, or download software. It is just one machine communicating with another one on a different network.

This reminds me of an old question that network engineers joke about. If you have to send a very large amount of data from Los Angeles to New York City to store the data in an archive, would it be faster to send it over the Internet, even with the fastest connections available, or to load a large cargo jet with solid-state drives and fly it there? I'll let you think about that for a minute.

The answer is most likely an airplane can get all the data from Los Angeles to New York faster than the Internet because that flight has higher throughput, but the latency for that connection is much higher than the Internet connection. We'll discuss in this section what these terms mean and help you understand how to purchase Internet services, including discussing how to purchase truly redundant connections to ensure business continuity under any circumstance.

CHOOSING A SERVICE TYPE

In some areas, especially in rural places, you may only have one option for high-speed Internet access. Connectivity is increasing, with heavy investment from governments worldwide, so the choices available are also increasing. Getting the right service for your company is important to successful operations.

As with most things, having a trusted advisor to help you choose the right type(s) of services is essential.

When deciding on a service type, there are three major benchmarks to consider.

- Speed, bandwidth, or throughput (often used interchangeably): This is the measurement of the maximum amount data that can be transmitted at any given time across the connection. Most people focus on this measure because it is what is generally advertised by the service providers. It is relatively easy to measure by going to speed test sites, such as fast.com, speedtest.net, or speed.measurementlab.net. If you are going to test your connection, it should be when a limited number of people are using the connection so that you aren't sharing the test with someone else's usage. This is usually measured in megabits per second (mb/s) or gigabits per second (gb/s). Higher speed is better; anything over 25 mb/s is considered broadband or high-speed Internet.

- Asymmetric or symmetric: Some types of Internet services have different speeds for upload and download usage. This means that you may have less ability to send data than to receive data. As companies are using cloud services more and more, the ability to send data is just as important as to receive it. Many companies are finding that using symmetric connections improves the quality of experience of cloud and VoIP services. Symmetric is a better connection because it allows you to send and receive at the same speeds, allowing modern applications to work efficiently.

- Latency: This is the measure of how long it takes for a standard packet of data to go from your connection to a location on the Internet and back. Services with lower latency have better quality of experience because most connections are very chatty, meaning that each packet of data that one side sends to the other is confirmed by the receiving side. The longer it takes for the packet to go from one location to another, the worse the quality of connection. Lower latency is better, as it allows the data to get from the sender to receiver faster. Latency is measured in milliseconds (ms). Any latency above 100 ms will cause interactive services such as VoIP or cloud desktop services to behave poorly, such as chopping phone calls, slow response from websites, and interactive applications to not work effectively.

Notice that I didn't mention the *type* of Internet service in the three considerations above. Once you have determined which factors are the most important to your company, you'll be able to decide which type of service you should get. For example, if you have a large office (one-hundred-plus employees), have mostly in-house servers, and use VoIP phone service, your two top priorities are speed and latency. If you have a medium-sized office (twenty-plus employees) that use mostly cloud services and VoIP phones, symmetric connection and latency are most important.

TYPES OF INTERNET CONNECTIONS

Now that you have an idea of the major factors in choosing an Internet connection, let's discuss which types of service are best for the major considerations, along with overall pricing comparisons between these services. There are always new types of services coming out, so this isn't a definitive list, and all of these may not be available in your area. We'll discuss these from highest price to lowest price.

INTERNET CONNECTION TYPE	DESCRIPTION
Metro Ethernet fiber or dedicated fiber	There are two distinct types of fiber-optic services. The best service is *dedicated fiber*. This type of service may have different names depending on the service provider, but generally, they'll understand you if you tell them one of these two terms. This service is the most expensive but has the highest available bandwidth (depending on your provider), usually is sold as a symmetric service, and has the lowest (best) latency. Speeds usually start at 100 mb/s up to 100,000 mb/s (100 gb/s). Since this connection is dedicated, you should always have the same speed or bandwidth. This service is best for companies that have mostly cloud services, one-hundred-plus employees in the office, or work with large files.
GPON, XGPON, XGSPON, or shared fiber	This is the second type of fiber service that has become common. It is the lower-cost option, which most homes and businesses subscribe to. It uses passive interconnections to deliver fiber-optic service at a lower cost rather than running dedicated fiber-optic connections to each subscriber. Depending on how a service provider configures the network, speeds can be from 10 mb/s to 10,000 mb/s (i.e., 10 gb/s). Latency can be higher on shared fiber, but it is usually still considered very good. Because this is a shared service, speeds can decrease, and latency increase if the service becomes congested in your area. This is generally the next best service to dedicated fiber, unless you are in a very congested area of your provider's network.
Cable modem	Cable companies started delivering Internet service over coaxial cable in the 1990s. Speed and latency have both improved significantly since the first deployment, with speeds now reaching 1 gb/s. Cable modem service is almost always asymmetrical, and it is shared service, which means that if the node (service connection point) in your area is congested, speeds and latency can suffer. Cable modem service is great for small businesses because it is generally lower cost and can offer high speed service in most areas.

INTERNET CONNECTION TYPE	DESCRIPTION
Cable modem	Cable companies started delivering Internet service over coaxial cable in the 1990s. Speed and latency have both improved significantly since the first deployment, with speeds now reaching 1 gb/s. Cable modem service is almost always asymmetrical, and it is shared service, which means that if the node (service connection point) in your area is congested, speeds and latency can suffer. Cable modem service is great for small businesses because it is generally lower cost and can offer high speed service in most areas.
Fixed wireless (FWA), microwave or millimeter wave	Internet service is delivered to your location over the air through wireless connections from a remote tower or another building. This is not to be confused with cellular or satellite Internet service. Fixed wireless can be either symmetric or asymmetric with speeds from 1 mb/s to 10 gb/s. Latency depends on the service connection type and if it is shared (also known as point to multipoint) or dedicated (point to point). Speeds and service can be as good as fiber, usually in urban areas where the service location is close to a tower site. Weather can affect fixed wireless, especially hard rainfall, heavy snow, or ice. FWA is popular for rural areas where the cost to build fiber is high. It is also a great option for true redundancy of service, which we'll cover in the next section.
Satellite	There has been a significant increase in the news about satellite Internet service with the recent advent of SpaceX's Starlink and Amazon's Kuiper. There are other options, such as ViaSat or HughesNet, that have been available for some time. This service is similar to fixed wireless, except that the "tower" locations are in space. Satellite service generally has high latency and varying performance depending on the location of the satellites, the amount of congestion on the network, and the weather. Satellite service is good for very rural locations where other service is not available.

INTERNET CONNECTION TYPE	DESCRIPTION
Cellular wireless	Many cellular companies are rolling out 5G service in urban areas, promising Internet speeds upward of 1 gb/s. Although great for mobile devices or as backup service, cellular service can vary in quality, greatly depending on the time of day and how many other devices are connected to the tower that you are using. Latency and jitter can be very high with cellular service. Finally, most cellular companies have a maximum amount of usage before the service provider starts to charge more for service or reduce the speeds delivered. Cellular wireless is a good backup option.
Digital subscriber line (DSL) and T1 service	I'll put both services in the "dishonorable" mention category since speeds are poor for both types of connections. Service is delivered over copper cables that the phone company has had in the ground for forty-plus years. T1 service used to be the gold standard for Internet connections in the 1990s and 2000s but has long been passed by because the maximum speed is 1.5 mb/s, hardly enough for any business today. DSL speeds are higher but usually capped at 25 mb/s. The further away your location is from one of the telephone company's central office locations, the poorer the service you will receive. Only buy this service if it's the last resort.

As you choose your Internet connection, ask the provider about their SLA. This gives you a guarantee of speed, availability of service, and latency. This is especially important when you are relying on the service for critical connections, such as if you are hosting your own servers for outside access or are relying on cloud services to run your business. You want to see what type of guarantees your provider is willing to put on their service to show you how reliable it really is.

True Redundancy

Even the best Internet connections go down periodically. Whether it is for maintenance, network outages, or line cuts, no Internet connection is 100 percent reliable. If your company absolutely needs your Internet connection to be operating all the time, having redundant connections is essential.

Quite a few years ago, one of my clients suffered a significant outage due to what I call "backhoe monsters." These are large yellow machines that munch on buried cables for dessert. I received a call from the client that their Internet and phones were down. I called their Internet provider, and they stated that they were showing an outage in the area and were dispatching repair crews. I decided to take a drive out to the location to see what was going on. There was a freeway construction project going on near their offices. The crews had not properly marked buried cables and cut the main feed from one side of the freeway to the other. All the service providers were using the same conduit to cross the freeway, so it took down service in the area for three days while they ran new cables. Luckily, we were able to set up a fixed wireless solution for the client to bring up their Internet while they were rebuilding their underground cables.

The following are a few things to consider when ordering redundant Internet connections:

- Redundant connections should be from different providers. One single provider could have a global outage, even if they are providing service to you in multiple ways.
- The connections should come into your building through different paths. Even if you order fiber-optic service and cable modem from different providers, they are most likely coming into your building through the same underground

conduits or across the same poles outside your building. Verify that the paths are different, or better yet, use fixed wireless, satellite, or cellular for one of the connections to provide better redundancy.

- If you have limited options in your area, develop a contingency plan if your Internet is offline, such as having your employees work remotely or some type of cellular hotspot service.

Combining Internet services into one building requires more than just switching cables from one connection to the other when service goes offline. Having the connection switchover happen automatically is important to keeping your service alive when one provider goes down. There are several options to provide this redundancy, which depends on the level of sophistication required for the services that you have running on your network.

AUTONOMOUS SYSTEMS

The original method to provide redundancy of Internet connections is rather complicated and generally used by large corporations or ISPs to connect to each other. It requires the organization to register for an autonomous system number and purchase a block of IP addresses. It also requires expensive routers and maintenance of them. If you are a large business, it makes sense to go down this path. For most other companies, the other two options are best.

SD-WAN

As router technology has improved, one option for true redundancy is to deploy software-defined wide area networks (SD-WAN). My recommended method for deploying this service entails ordering services from multiple service providers (fiber, cable, fixed wireless, cellular) and then signing up for service with a SD-WAN provider. Some ISPs sell SD-WAN services directly, but this puts all your eggs in their basket because they are providing all the connections. If that ISP has a network-wide outage, all your services will go offline. The SD-WAN provider will install a router at your office(s), and you plug all your Internet connections into that device. The SD-WAN router then will do some magic, which includes advanced wizardry such as detecting which connection has the best performance for high-priority service such as VoIP, ensuring that hosted services are online, and reducing speeds for services that aren't as important, such as video streaming. The router will also detect if one of the connections goes offline or has data loss and automatically switch to the other connection(s). The SD-WAN connections can also provide private connections between locations (similar to a VPN) so that remote offices can access a corporate network.

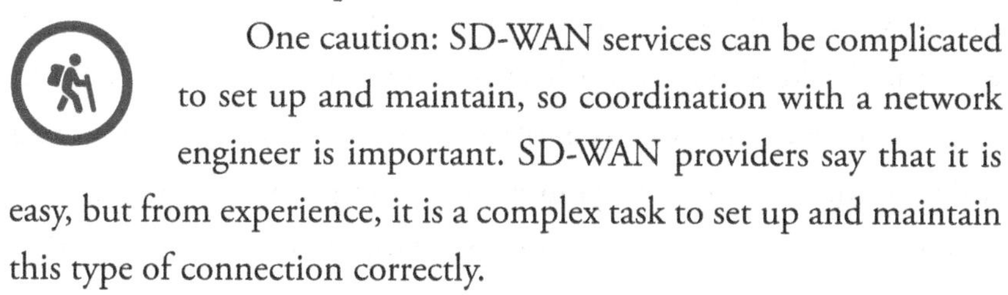

One caution: SD-WAN services can be complicated to set up and maintain, so coordination with a network engineer is important. SD-WAN providers say that it is easy, but from experience, it is a complex task to set up and maintain this type of connection correctly.

ROUTER FAILOVER

Most business routers have the ability to failover from a primary connection to a secondary connection in the event that the primary service goes offline for any reason. The advantage to this method of redundancy is it is the simplest to set up and maintain. There are two downsides.

- You will only be using one connection at a time. Your backup connection generally will not be used unless the primary service goes offline.
- If you host any servers, VPNs, or any outside connections to the network, these services will most likely not be available when the primary connection goes offline (at least temporarily).

Whichever method your company decides to implement, having redundancy of network connections has become essential for most businesses, even with hybrid or remote work. Having a secondary connection with true redundancy is an inexpensive insurance policy against Internet outages.

Operating after a Failure

What happens when everything you've implemented to protect you and your company fail? We've reviewed together a vast number of ways that hackers can get into your systems and cause damage. Even with all the right protections in place, there is a good chance that you will wake up one day with your systems compromised and not able to function normally. There are two backstops that I recommend that you put in place. The first is actual insurance on your company. The second is a business continuity system. I highly recommend that you work with seasoned professionals to implement these. Let's dive into both recommendations to give you an overview.

BUSINESS INSURANCE POLICIES

Beyond normal business insurance, there are two types of insurance policies that I recommend that every company maintains. General liability coverage isn't sufficient to protect you from all the threats to your business. Take the time to work with your insurance broker to choose the right coverage for your company. Discuss the potential threats to your business, and make sure that your policies will cover them. You can run through a few scenarios that we've discussed in this book and understand how the insurance company will assist you in protecting your business from loss.

- Cyber liability insurance: This insurance is designed to protect you against external threats to the company, such as data breaches, ransomware, and other cybersecurity issues. You will be required to fill out an annual questionnaire about your operations. Depending on the size of your business and the level of coverage, you may be required to implement many of the security measures that we've discussed in this book. Required items include data encryption, employee training, multifactor authentication, and EDR/MDR systems. The insurance will help you recover your data, pay for system downtime, and work with an outside security company to find the method of entry into your system. It can sometimes pay for any fines, reputation damage, and lost business.
- Employee theft and crime insurance: This coverage will ensure you against internal threats, such as employee mistakes and misconduct. It may seem that this is only for employee's intentional acts against the company, but it also protects against errors. For example, if an employee falls for a social engineering attack and inadvertently wires funds to a hacker, this coverage will help recover the lost money. This also protects against an employee's misconduct that may cause losses for third parties, such as your vendors or clients.

Your insurance carrier will require you to be responsible in your actions and take the necessary steps to protect against the threats. You can't just have insurance and operate recklessly. Just like having too many accidents and speeding tickets in your car, insurance companies will not cover companies that don't take the threat seriously.

BUSINESS CONTINUITY SYSTEMS

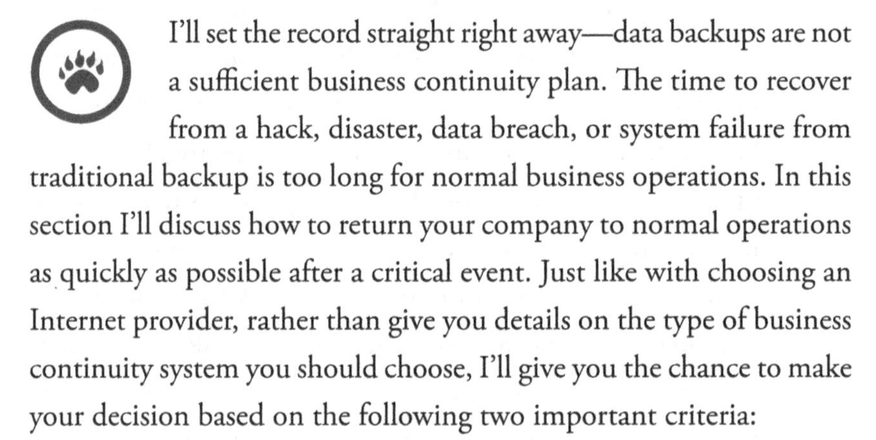

I'll set the record straight right away—data backups are not a sufficient business continuity plan. The time to recover from a hack, disaster, data breach, or system failure from traditional backup is too long for normal business operations. In this section I'll discuss how to return your company to normal operations as quickly as possible after a critical event. Just like with choosing an Internet provider, rather than give you details on the type of business continuity system you should choose, I'll give you the chance to make your decision based on the following two important criteria:

- Recovery point objective (RPO): This is the maximum amount of time between the system failure and your last successful backup of your systems (snapshot). For example, if your snapshots occur once a day at 2:00 a.m. and you have a system failure at 1:59 a.m., you have potentially lost twenty-four hours of data. This means that your RPO is twenty-four hours. To lower your RPO, you will need to increase the frequency of your backups.

- Recovery time objective (RTO): This is the maximum amount of time between a system failure and recovering to normal operations. As an example, let's assume you have one server in your office and are running cloud-based data backups. These backups are file-only backups, which means that they need

> to be copied back to a server in your office to operate again. You have a major thunderstorm in your area overnight, and your server received a catastrophic power surge at 10:00 p.m. Luckily, you have a warranty on your server, but the manufacturer needs to order replacement parts. They arrive the next day, and it takes another day to reinstall the server, download the software from backups, and bring the system back online. Your RTO is approximately sixty hours (10:00 p.m. on day zero to 10:00 a.m. on day three).

Many business owners that I talk to think that because they have backup systems in place, they will not have any lost data or lost time in the event of a catastrophic failure of their systems. But this is not the case! Every business continuity system has an RPO and RTO, and it is important to define those when choosing a continuity system.

The lower your RPO and RTO, the more expensive the system will be.

If you want near zero lost data and instantaneous recovery in most scenarios, your technical team will need to purchase large on-site and off-premises systems with advanced interconnections between them and constant replication of data between all your redundant systems. These types of business continuity systems are expensive to install and expensive to maintain. For most businesses, you must compromise the costs against some downtime and lost data. Meaning that for most businesses, in the event of a significant disaster, there will be some downtime and some portion of data that is lost.

Let's assume that your business continuity system takes snapshots of your primary servers every three hours and uploads those snapshots to cloud storage once per day at 2:00 a.m. You choose this schedule because you don't have sufficient redundant storage on-site to store more than eight snapshots in a twenty-four-hour period. If your servers, including the on-site backup systems, suffered a catastrophic failure at 10:00 p.m., your last successful backup to the cloud occurred at 2:00 p.m. the night before. This means that you have lost about twenty hours of data. Additionally, since your on-site systems are all offline, the time to recovery may increase to several hours or longer because of the time to connect to the cloud storage. Although this scenario is detrimental to the organization, your business would most likely be back in operation the same day with only one day's lost data.

Choosing a System

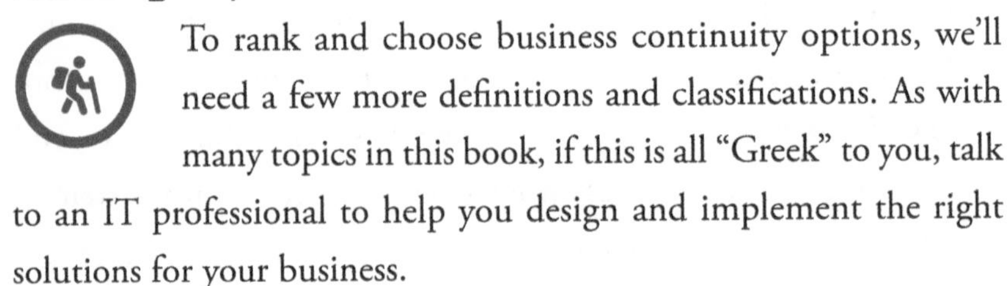

To rank and choose business continuity options, we'll need a few more definitions and classifications. As with many topics in this book, if this is all "Greek" to you, talk to an IT professional to help you design and implement the right solutions for your business.

- Hot standby: A system that is replicating everything that is happening on the live system. In the event of a failure of the primary system, the hot standby can take over operations nearly instantaneously. This is the fastest option for recovery.
- Warm standby: A system that is replicating everything that is happening at periodic intervals. In the event of a failure of the primary system, the warm standby can take over operations in a very short period. This is a midrange option for recovery.
- Cold standby: A system that copies data from the primary system periodically. In the event of a failure of the primary system, the cold standby must be started and potentially have data restored. This is the longest option for recovery.

- Disparate systems: The primary and backup systems are fundamentally distinct from each other. A cyberattack, a network failure, or other catastrophe on one system cannot affect the other system.
- Data storage: Where backup data is stored by the business continuity system.
- Retention period: The length of time that data is stored and can be recovered.

Types of Business Continuity Systems

Of course, all businesses would want to implement disparate hot standby systems with high retention periods. As I mentioned above, these types of systems are very expensive and difficult to maintain. Although not an exhaustive list, this generalizes the types of options that are available to most businesses. These are ranked with the best type of system at the top.

- On-premises virtualized servers with auto-failover and off-site replication: The primary two vendors to implement this type of solution is VMware VCenter and Microsoft Hyper-V. In the event of any type of system failure, exact duplicates of the servers are kept in redundant storage on-site and also replicated to an off-site location (either public or private cloud storage). This solution provides hot standby on-site and disparate storage in the cloud, assuring that there is protection between the on-premises and cloud storage. It provides the lowest RPO and RTO. In the event that the primary location is completely offline, the business can continue to operate from the secondary off-site location with little change to business operations.

- Business continuity appliance with off-site replication: An on-site server or set of servers replicates snapshots periodically from your primary systems. Then these systems periodically upload the snapshots to cloud storage for protection of on-site failures. This is considered a warm standby solution and can have disparate storage, depending on how it is configured. It provides good RPO and RTO depending on how it is configured, with recovery usually within an hour. In the event of a complete failure of the primary location, the off-site location can run the business, but it may be slower or not as efficient as the primary systems.
- Standby server with data backup: This is where you have a duplicate server or system available to put in place if the primary server fails. What I generally see is companies keep an older server or system after an upgrade that can take the place of the primary server temporarily in the event of a failure. Data is backed up on-site (not recommended) or off-site (recommended). In the event of a failure, the standby server has to be brought up to date with the current configuration and data. Recovery time can be within a day or possibly longer depending on the time it takes to get your backup data from storage.
- Data backup only: You have no duplicate server but store data on another type of media, such as tape (I still see tapes around today. Gasp!), hard drive, or cloud storage solutions. The time to recovery depends on the availability of obtaining a repair or replacement to the primary systems. RTO may be measured in weeks. Because of this, I do not recommend data backup

only. It can be used for retaining backup data for long periods of time but is not a business continuity solution.

There are many vendors that offer these types of services. When choosing a solution, make sure you ask and understand the RTO and RPO of the solution. Also, they should always be disparate systems because a common practice for hackers is to delete all backups that they can reach so that you will have to pay the ransom to recover your data. Some companies are paying the ransom even if they have valid backups because the time to recovery is too long.[24] According to Steve Long, CEO of Hancock Health, "The affected files were backed up and could have been recovered but restoring them would take days—maybe even weeks—and would be costly." If you have separate business continuity solutions, you can tell the hackers where to put their demands (#@!) and continue to operate your businesses normally.

Testing

The final step to implementing a business continuity solution is to test it regularly. First, your IT team should come up with a procedure to ensure the following two things:

- Success of data backup or replication
- Test recovery from system failure

Your IT team should receive notification of the success and failure of each job to replicate data. Successful replication should be looked at periodically (once a month is generally sufficient) to ensure that the reports are accurate. Any failures should be resolved immediately.

---

24 Samm Quinn, *Hospital pays $55,000 ransom; no patient data stolen*, January 16, 2018, http://www.greenfieldreporter.com/2018/01/16/01162018dr_hancock_health_pays_ransom/ (accessed June 6, 2022).

Second, your IT team should implement periodic testing of the systems. Part of the test should be to fail your primary systems and make sure that everything works as planned for the failover. This should include ensuring that your RTO and RPO are met successfully. Your systems should be tested by security professionals to ensure that a hacker is not able to infect, encrypt, or delete your secondary systems or backup data. This type of testing can occur often but should happen at least once a year or after any major changes to your network.

Having the right business continuity systems in place is the last line of defense against attacks on your company.

Having the right business continuity systems in place is the last line of defense against attacks on your company. The hope is that you will never have to use them; however, the reality is that someday when disaster strikes, you will be happy that you had them in place.

Network Design and Business Continuity Checkup

Give yourself a score of 1 to 3 in each area. Add them up to see how you did overall in this area. Focus on just one area to improve upon for now.

AREA OF CONCERN	SCORE
I have implemented redundant Internet connections at my business locations where service is critical.	
My company has sufficient cyber liability and employee crime insurance to cover any potential losses.	
I have implemented the systems and procedures required by my cyber insurance company.	
The business continuity systems I have in place align with my desired RPO and RTO.	
My IT team tests our business continuity solutions regularly to ensure that they will work as expected in a real disaster.	
TOTAL **TOTAL POSSIBLE 15**	

NOTES

Network Design and Business Continuity Key Points

- Businesses rely on the Internet to conduct operations.
- All Internet connections will have downtime. Implementing redundant connections will avoid business operation interruptions.
- Having redundant Internet service is an inexpensive insurance policy.
- Choosing the right Internet service for your business requires some additional analysis and questioning of the service provider.
- Cyber liability and employee theft insurance is essential for business operations.
- Data backup alone is not a business continuity solution.
- Choosing the right business continuity system should be based on your required RPO and RTO.
- There are many options available for business continuity systems, and relying on an IT professional is important to enable you to choose the right system.

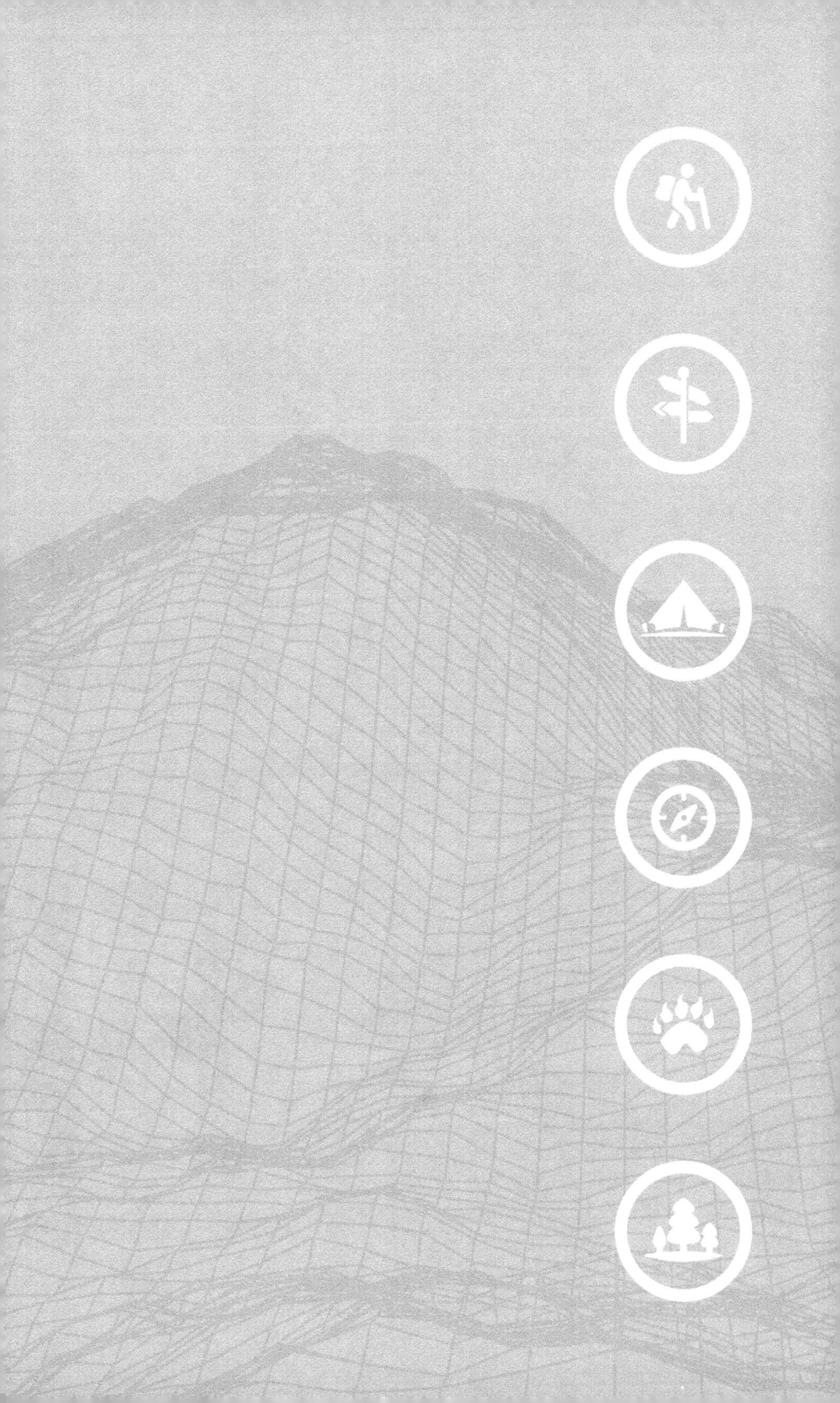

CHAPTER 9

WORKFORCE MANAGEMENT—HYBRID, REMOTE, AND OUTSOURCED WORKFORCES

When initially planning this book, I had thought this chapter would be a small part of a larger chapter on business technology strategy. After the events of 2020, that focus has completely changed. In March 2020 most states ordered companies to send the majority of their workers home. Over the next month, businesses frantically figured out how to continue operating effectively with most employees working from home. They also competed with their employees' children attending home school in the same physical space and using the same Internet bandwidth. This put a strain on families, employers, the Internet, *and* security professionals. In this chapter we'll discuss things companies did to be successful in the transition, tools they used initially, and how businesses can now use these products and services in the future to develop hybrid work environments for their employees.

One thing is completely clear: very few businesses were prepared for the dramatic change that occurred so suddenly. Medium and large companies usually develop disaster plans that consider natural disasters, terrorist attacks, or other dramatic events that change the local business environment temporarily. Very few planned for a pandemic that would send the whole world into lockdown for an extended (and possibly indefinite) period of time.

For IT managers, having all employees working remotely without a plan is a nightmare.

For IT managers, having all employees working remotely without a plan is a nightmare. Rather than protecting one network, with all the layers of protection from firewalls, intrusion detection systems, and physical security, each remote worker also has their own home network. Even worse, many remote workers were told to use their personal computers to connect to corporate networks. That brought in the possibility of a home PC infected with a virus, ransomware, or other malware connecting into a business network and infecting all the other computers connected. It also meant that IT professionals would have to support home-based Wi-Fi, routers, and networks. It dramatically increased the amount of work required to support the IT infrastructure.

Amazingly, many companies implemented plans very quickly and got their employees back to work. Businesses that embraced technologies such as cloud services, VoIP phones, and virtualization fared much better than those that didn't. New technologies also emerged that made remote work even better. Luckily, technology did allow for the transition to happen for many workers. If the COVID-19 pandemic had happened even five years prior, the situation would have been far worse. In a sense we were lucky that technology was available to keep companies operational.

At my company I remember distinctly the week that the world changed. We hold our management meetings every Thursday. On March 5, 2020, we were discussing our response to the increasing pandemic. At that point there was no stay-at-home order in our state. We decided out of significant caution to send all employees home starting on Friday. We discussed how that would work for our staff because most had desktop PCs rather than laptops. For most employees the applications that we use are cloud-based, so transitioning away from the office was relatively simple. For the accounting team, they needed access to our office server.

We decided to send most employees home with their office computers. This allowed them to transition easily to work from home. As a VoIP provider, it was easy for our users to either take their VoIP phones home or use a "soft phone," which is software that runs on the computer that acts as a phone. For users who still needed access to data at the office, we had both a VPN connection available and a Remote Desktop server for remote access. We had recently been using Microsoft Teams in the office for chat, which we still rely on extensively for all intracompany communications. As of Friday afternoon, we were almost fully operational remotely.

That Sunday, the governor of our state announced a stay-at-home order, which sent most workers home. As an IT service provider, our clients relied on my company to help them make a successful transition. We had only one day to figure out how to make it work, so *obviously*, we were fully experienced in remote workforce management. We offered our clients three options for their employees to work remotely: use the same tool that we use to provide remote support to our clients to access their own machines, set up VPN access, or if they already had it available, use remote desktop services. We didn't have time to configure anything new, so we went with the tools already available. Amazingly, we were able to help most clients transition easily to remote work. There were instances where companies were not allowed to operate at all, such as retail, dental offices, and entertainment and hospitality (which is a big industry in one of our company's locations). In those cases, we locked down their company's network and helped them pause most operations until they could start back up again.

Several years later we've learned numerous lessons about effective remote workforce management, both from how to deploy technology and how to manage people through technology. Because remote and hybrid work is here to stay, I'll discuss both in this chapter so you can be successful in this new frontier of work.

Key Definitions

- **Brick-and-mortar business:** The company has traditional offices, retail space, or manufacturing facilities where all employees are expected to work.
- **Remote workforce:** All or the majority of the company's employees do not work at a company facility. The employees work from home, from their own facilities, from the beach, or on a spaceship. Anywhere but a company location.
- **Hybrid workforce:** Company employees have the option to work either remotely or from a company location.
- **Work from home:** Temporarily working from a remote location for personal or company reasons.
- **Blended teams:** A company team has a combination of office, remote, and/or hybrid workforce.
- **Dispersed teams:** Employees are spread out between many company locations and are expected to collaborate and work together.
- **Hot-desk:** Shared workspace that allows an employee to work from the company's offices at any desk they choose.
- **Asynchronous communication:** A type of communication where immediate response is not expected. This may include email threads, communication boards, or other ways to communicate with an individual or group of employees that does not expect immediate response.

- **Synchronous communication:** A type of communication that is interactive, and a response is immediately expected. For example, chat or messaging programs, text messaging, or phone calls.
- **Real-time communication systems:** Software or applications that allow teams to communicate instantaneously with each other through text, audio, and/or video.
- **Productivity:** The measurement of effectiveness of work in number of units per a specific time period. For example, in manufacturing, how many widgets are produced per hour in a factory.
- **Knowledge worker:** An employee primarily interacts with knowledge and information rather than physical objects or materials. They usually focus on problem-solving or strategic tasks.
- **Flex time:** Employees are given the ability to choose their hours of work so long as the tasks and assignments are completed and results are achieved.
- **Offshoring:** Workers are based outside of their primary country, usually at a distant location and in dramatically different time zones. For example, in the United States, an offshore employee would be located in the Philippines or India.
- **Nearshoring:** Workers are based outside of their primary country, usually at a relatively close location and similar time zones. For example, in the United States, a nearshore employee would be located in Central or South America.

- **Cloud computing:** A system that allows data to be accessed from nearly any location so long as both systems are connected to the same network (such as the Internet).
- **Public cloud:** A company uses a shared service, such as Microsoft's Azure or Amazon's AWS, to store their data and operate their systems.
- **Private cloud:** The company owns and operates its own systems to allow employees to access the company's services from any location.
- **Hybrid cloud:** Data and services are spread between public and private cloud services.
- **Voice over Internet Protocol (VoIP) phone systems:** Phone systems that connect directly to a network and operate as service over the Internet.

The Need to Change

A global pandemic is just one reason why companies are having to make a shift away from brick-and-mortar business practices. The pressure on businesses to offer their employees more flexibility in work locations and schedules has been brewing for many years; the pandemic just pushed the pressure over the edge from want to need. Companies that do not offer flexibility are going to have a challenge attracting and retaining top talent. According to Steven Rothberg, president and founder of CollegeRecruiter.com, there will be "tremendous upheaval over the coming months with many employers unable to attract enough workers to keep their doors open. It isn't that people won't want to work. It is that they won't want to work for those employers."[25]

There are instances where remote work for all employees is impossible, such as in retail, manufacturing, and direct service environments. In these scenarios, production and customer-facing positions may still be required to be in a company facility or office, but the backend support employees can work remotely, thus creating a blended team.

Finding talent is another strong argument for flexibility in the workforce. Workers have shifted to living where they want rather than where they work. Additionally, only looking for employees in a single location dramatically narrows the total pool of possible candidates. Opening the search to locations other than where your office is located will allow you to find employees who may be more educated, talented, or dedicated to your company. It also allows companies to offshore or

25 Joe McKendrick, *Remote And Hybrid Work Is Here To Stay, And That's Why Quality Of Worklife Matters*, August 31, 2021, https://www.forbes.com/sites/joemckendrick/2021/08/31/remote-and-hybrid-work-is-here-to-stay-and-thats-why-quality-of-worklife-matters/?sh=118793427091 (accessed June 14, 2022).

nearshore positions when no candidates are available in your area, or the wages have become too expensive for your continued operations.

There are significant disadvantages to remote and hybrid work. Employers are often concerned about productivity. Many owners have come to me asking for help to make sure their staff is producing results when they can't just walk around the office and check on everyone. Team cohesion and personal connection are also big challenges. From my experience the lack of physical proximity seems to cause more stress and can lower overall morale. Onboarding, training, communicating, and discipline are also big challenges. There are tools that have been developed to overcome these difficulties, which we will discuss in this chapter. These systems allow managers and owners to lead their teams effectively wherever they are located. New tools are always being developed, so my intention is to give you some ideas to allow you to find what is right for your company and to be successful in this new frontier of work.

Moving to the Cloud

I used to say that cloud computing is just using someone else's servers, meaning that instead of making the capital investment in your own hardware, you rent servers from another company. The cloud has become much more than that and is now integral in allowing companies to move to remote and hybrid work. Whether it is private cloud (meaning you still own the hardware) or public cloud (meaning you are using another company's servers and management), cloud computing allows companies to be more flexible, allows their users to connect from anywhere, and scale resources based on the demand of their employees. There is much more to cloud computing than fluffy white pillows in the sky. Like many of the topics I've covered in this book, the cloud is often misunderstood because it's complicated. Sadly, it is often used as a marketing word by companies trying to sell you services. I'll take you on a little journey through the sky to find the pot of gold at the end of the rainbow. I'll also try to stop using puns to make the material more interesting because I know I'm the only one laughing at them!

We'll focus on the following four areas to help your employees work effectively anywhere they are (company's office, home, hotel, or a random undisclosed diner). My goal is not to give you a list of products to buy; it is to introduce you to technology that makes hybrid work easier.

- Communication
- Collaboration
- Virtual Desktop
- Team Engagement

COMMUNICATION

Over the years of running my company, I've listened to many business speakers and coaches about managing and leading employees. One of the common recommendations that I've heard is "management by wandering around," introduced by Tom Peters in his book *In Search of Excellence*.[26] In other words walk around and ask how the team members are doing. Listen to their conversations. Engage in the work that they are doing. Ask them to help you solve problems.

---

26 Thomas J. Peters, *In Search of Excellence: Lessons from America's Best-Run Companies*, (New York: Harper Business, 2006).

Since moving to hybrid work inside my company and having employees across the country, I've often asked myself if I am doing enough to wander the office and engage with my team. Although it may not be initially intuitive, there are many tools to allow your teams to maintain communication even if they are not in the same physical location. It may not be exactly the same as wandering around the office, but it can be just as effective in my experience.

There are two types of business communication: asynchronous and synchronous. Asynchronous communication is based on technology that does not expect an immediate response from the other party. Examples include leaving voice mail, email, physical mail, and packages. Technology has existed for many years to allow for asynchronous communication. Synchronous communication anticipates an immediate or near immediate response. Some examples include phone calls, text messages, instant messaging, and video conferencing. Successful remote or hybrid businesses use a mix of both and identify when each type of communication should be used.

Although not a complete list and noting that products and technologies change very quickly, here is a list of common tools and systems that are used commonly in businesses today for communication.

Asynchronous Communication

SOFTWARE OR SYSTEM	PURPOSE	BEST USE
Microsoft 365 Email	Business communication and electronic mail system.	Most small- to large-sized businesses, especially those that need security, mobility, and synchronization.
Google Workspace Email	Business communication and electronic mail system.	Small- to midsize businesses that don't use Microsoft Windows as the primary operating system.
Other Email Systems (Gmail, Outlook.com, Yahoo Email, IMAP, POP3)	Micro-business and individual electronic mail system.	Sole proprietorship or personal email communication.
Post Office Mail	Old-fashioned letter communication.	Because of all the electronic communication today, sending a personalized letter, card or note makes a big impact!
Overnight Packages	Sending important documents quickly.	Usually used to send important documents or packages. Can also make a bigger impact than a letter to a customer or prospect.
Voice Mail	Let the receiver know why you picked up the phone to call, and request a call back.	Why not leave a voice mail? Let the person know who you are and why you called.

Synchronous Communication

SOFTWARE OR SYSTEM	PURPOSE	BEST USE
Phone Calls	Reach out and talk to someone.	Because of the amount of spam phone calls, many may not pick up the phone, but it is a tried-and-true way to get hold of someone quickly. For millennials, send a text first and then call.
Microsoft Teams	Instant messaging, group communication, file sharing, and video conferencing.	Most small to large-sized businesses. Has the ability to archive all data for enterprise requirements.
Google Chat	Instant messaging and video conferencing.	Small business communication.
Slack	Instant messaging, group communication, file sharing, and video conferencing.	Working between organizations and community groups. Small and midsize organizations.
SMS Text Messaging	Quick communication with employees and clients.	Send messages between team members. Allow customers and clients to send messages and requests to the company. Send notifications to clients about upcoming service and appointments.
Zoom, GoTo Meeting, or WebEx	Video conferencing and webinars.	Communication to feel like you are as close to being in person as possible.
VR Workrooms	Virtual reality meetings.	Give a more realistic virtual meeting by using virtual reality headsets to feel like you are together.

One of the big faults I see in my own business is that the wrong type of communication medium is used. For example, people send emails and often expect an immediate response. The person on the other side of the email may not respond for hours because they are working on another project, meeting with clients, or just have focus time during which they don't look at their emails. Use a synchronous communication method to get hold of them quickly. On the other hand, if the communication requires detail or sharing of information, sending a text message is inappropriate. Have you ever received a text that is paragraphs long? It is frustrating because you must then turn your attention to a "novel" and not a short message. Choosing the appropriate method of communication, whether in business or personal communication, is very important to allow the conversation to be productive.

I've found that companies that have been able to effectively use the right communication among their teams are more successful in remote or hybrid environments. The effective use of team meetings, such as daily huddles (quick gatherings among some teams), weekly one-to-ones (thirty-minute manager-to-employee discussions), employee virtual social hours, regularly scheduled management meetings, and company-wide staff meetings, is an effective strategy to allow teams to communicate as if they are physically in the office. Also, I've seen success with making these meetings enjoyable with games, get-to-know-you questions, and employee involvement in planning and holding them. Team members respond differently to the different meeting styles, so varying them is important to getting all involved.

One final note on communication: It is important that your team has a way to notify others when they will not be available to respond. Having out-of-office responders set, focus-time notifications on synchronous communication platforms, and shared calendars for effective meeting coordination all assist with helping team members not feel overwhelmed with too much communication and too many meetings. Finding the right balance is essential for productivity among hybrid and remote teams.

COLLABORATION

Another difficult part of working as remote teams is collaborating on projects and tasks. In an office environment, a team could go into a conference room, with a whiteboard, paper, and many ideas. Brainstorming, discussion, and work could happen interactively in real time. It is difficult to replicate that interactive effort in a virtual environment. What I have seen is that when using the right tools, the interaction can be better and more effective than even before. Teams can get together virtually and work together on projects, documents, and tasks in real time, getting real work accomplished quickly and efficiently. What may be lacking in the physical element is made up for with the virtual tools.

Google was one of the first companies that developed technology that allowed multiple people to interact on a document together at the same time. The technology was originally called Google Apps and has evolved into Google Workspaces. The first time I saw two people working on the same spreadsheet in real time at distant locations, I was very impressed. You could see where the other person was in the file and the changes they were making in real time, and they could see what you were doing to the file. Prior to this technology, when one user opened a document (usually stored on a file server), the file was locked, and another user could only open it in read-only mode. Another user could not make any changes until the file was closed and saved by the other user. Did you ever have that experience when you needed to get into a file, and someone had left it open overnight and didn't close out of it? So frustrating! With this new technology, that would never be an issue.

Other systems have come along that have allowed similar collaboration with tools that we are familiar with. Microsoft, not to be outdone, has incorporated similar technology in their Office Suite, along with OneDrive for Business and Teams to allow the same interaction. OneDrive for Business, also known as SharePoint, allows companies to store and manage files, internal websites, and databases online. With this system, users can access the files and data across all their desktops and mobile devices. System administrators can set up the same type of access permissions that they could with in-office file servers. It makes mobility simple, with the users being able to access the data wherever and whenever they want.

In my company, we use Microsoft tools, and I'll give you an example of how we use them.

The team lead will create an online Teams meeting invitation. At the appropriate time, the group joins the meeting. The team lead will send an agenda to the group using the chat function, and everyone can open that document, even if they are using their mobile phone or tablet. Anyone in the meeting can add or make changes to the agenda. Everyone else can see those changes in real time. It is a group effort to create and manage the agenda. Rather than having static paper copies of the agenda that is created long before the meeting, it becomes a living document. Throughout the meeting, we take notes, add to-dos and action items, and mark off agenda items that are completed. If we are working on other documents, we can all open those files on our own computers, make changes in real time, and see what others are adding or changing also. We can also share screens so that we can see the work that someone else is doing. All of this is done with everyone on video so we can see and interact with each other. We've found that the meetings are far more productive than our in-person group meetings because of the real-time interaction that occurs seamlessly.

The big part that is lost in these virtual meetings is the small talk. I've found that even in relatively small groups and especially in large groups, there tends to be little interaction before or after the meetings. It makes them more efficient, but the teams don't bond as easily. I have two tricks that have helped me.

Trick Number 1: Have a *segue* before each meeting. I'm shamelessly stealing this idea from *Traction*.[27] What is a segue? Simply, each person in the meeting has to give one quick good news of the week, either from their personal or business lives. It helps open up the meeting and create a little interaction beforehand. It's much better than staring at each other in virtual land or watching the other people check emails or text messages. I've found that it allows the ideas to flow much more easily through the meeting.

Trick Number 2: Do breakout rooms. This is a relatively new technology incorporated in most virtual platforms such as Teams and Zoom. I'll ask a question or assign a topic and split the group up into smaller groups, usually five people or less per breakout room. People are much more comfortable talking to a smaller group than speaking up at a large meeting, even if that meeting is only ten people. I'll give the breakout rooms a certain amount of time to collaborate and then come back together. Each group will have a spokesman present to the larger group. It is very effective to get people to present their ideas and concepts.

Using modern tools such as Microsoft 365 and Google Workspaces has allowed companies to work even more efficiently in a virtual

---

27 Gino Wickman, *Traction: Get a Grip on Your Business*, (Dallas, Texas: BenBella Books, 2012).

environment than they were previously. Many companies, including ours, have virtual meetings, even when they are in the same office building because the real-time work that can be completed through these technologies can make the meetings much more effective.

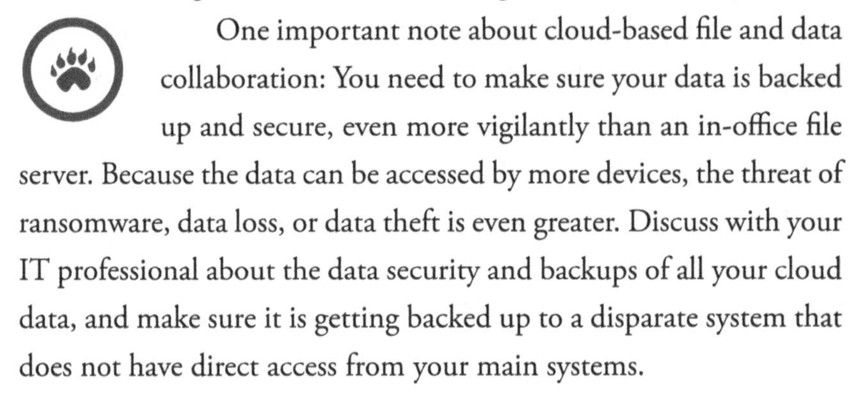

One important note about cloud-based file and data collaboration: You need to make sure your data is backed up and secure, even more vigilantly than an in-office file server. Because the data can be accessed by more devices, the threat of ransomware, data loss, or data theft is even greater. Discuss with your IT professional about the data security and backups of all your cloud data, and make sure it is getting backed up to a disparate system that does not have direct access from your main systems.

VIRTUAL DESKTOP

Even with online communication and collaboration tools, companies need to be able to run line-of-business applications, such as accounting, engineering, production, or legacy applications that are client-server based. That means a file or database server is required and the application runs on a desktop PC that accesses the servers. In a full-cloud remote access environment, running these applications remotely is slow and inefficient.

Microsoft released Windows Virtual Desktop (WVD) to the general public in March 2020 (perfect timing for the pandemic!) as part of their restructuring of their Microsoft 365 services. Rather than a revolutionary product, it is an evolutionary product that combines several existing product lines in a complete package. These include their legacy Office 365 (now Microsoft 365) suite of office tools, Microsoft Azure cloud services, and a product called FSLogix that glues the two together. For most businesses, using a brand-new, bleeding-edge, revolutionary product can be scary. WVD is a suite of tried-and-true products used by large and small businesses for many years, now packaged in a complete suite that small businesses can use to their advantage in the marketplace.

What WVD Is Not

We know that when techies start talking about their new "fang dangled" technology products, most people's eyes glaze over. Before we get into the juicy details about what WVD is, I want to dispel a few myths you may have heard about cloud computing and virtual desktops.

- It is not *warp drive* for your starship, although it will accelerate your business past your competitors running on old servers sitting in their dusty closet.
- It won't take you on the *Kessel Run in 12 parsecs*, but it will allow your employees to access their desktop, programs, and files wherever they are, even from a web browser.
- It is not the *Golden Idol* from an archaeological dig; however, IT professionals have been searching for this complete solution for many years.
- Your data does not exist in a *Cloud*, miles above the earth. Your data is stored in secure data centers across the country operated by Microsoft.

In nontechno speak, what is WVD?

Microsoft Azure allows you to run servers and computers in a data center, which is a building with redundant power, Internet connections, and physical security that is managed by Microsoft. Rather than having to buy new servers and computers, you rent them from Microsoft. WVD is a technology that allows your employees to access their desktop that is running in Azure from anywhere they have access to the Internet. WVD solves the technology problems businesses are facing in a hybrid or fully remote workforce.

Flexibility: You can access WVD and Microsoft 365 from pretty much any device from anywhere. Desktop, laptop, tablet, cellular phone, or web browser—access your files, documents, programs, and Windows desktops anywhere and anytime.

Security: WVD is built on top of Microsoft Azure. Microsoft won the award from the US Department of Defense for its JEDI Top Secret cloud solution.[28] It is the most secure cloud environment available. Security in Azure is integrated and includes multifactor authentication, user policies, backup, and retention policies.

Scalability: With WVD, you can add new users and remove users whenever you want. You can add extra storage, new servers, and deploy new programs quickly without buying more hardware. This allows your company to change as the business environment changes (for better or worse). When users aren't logged in and using their systems, their desktops shut down, saving you money.

Performance: The server infrastructure in Azure is modern and scalable. Your data is stored on high-performance SSDs. You can assign users to shared, dedicated, or high-performance graphic desktops, all depending on their needs.

---

28 Mark Haranas, *Microsoft Azure Creates Top Secret Government Cloud As Jedi Battle Rages On,* December 07, 2020, https://www.crn.com/news/cloud/microsoft-azure-creates-top-secret-government-cloud-as-jedi-battle-rages-on (accessed July 16, 2022).

Reliability: Your data is spread across multiple servers in highly redundant data centers. You no longer have to worry about the power being on at your office, your Internet service connection being cut, or any other variables that plague computer infrastructure in your office. If you have issues at your primary business location, you can send your employees to work wherever it is convenient for them.

As with any complex system, having a team of professionals who are certified in these types of environments makes running them much more efficient and secure. Discuss this option with your IT professionals, and find a company that can assist you with setting up and managing a cloud solution for your company.

TEAM ENGAGEMENT

Before the COVID-19 pandemic, most business owners that I know struggled with remote workers for two reasons: potential lack of productivity and difficulty with team engagement. We've discussed the first issue already and have shown that with the right tools, productivity can be increased with a remote or hybrid team. Team engagement or, more broadly, company culture, is more difficult to maintain with a separated workforce.

"Of course, it's not that company culture somehow goes away in a remote or hybrid context. Cultural beliefs and norms are still being created and reinforced, but they're not being guided by systems and routines that were previously established in the office. They're more open to change and subject to influence from new, non-work factors present in employees' day-to-day lives."[29]

I believe that company and team leaders must be proactive with developing and maintaining intentional culture when employees are not in the office every day.

There are several areas where technology can assist leaders in providing team engagement and maintaining company culture even with remote employees. I have had success with these ideas in my business, and I've seen them succeed in others. As with any new system, you need to make sure that you are solving a problem or enhancing your company with the new software. Implementing technology isn't just about the technology; it is about the results that it brings to your organization.

Human Resource Management

Small businesses generally don't have the HR systems that large companies have created. Most small businesses have one person assigned, such as the office manager, bookkeeper, or even the owner, who handles onboarding, payroll, and HR questions. In an office environment, it is easy for an employee to walk into the assigned person's office and get their needs taken care of. Employees can feel lost if they don't know who to go to or how to handle HR questions.

There are several good virtual HR systems that allow small businesses to provide a complete HR platform for employees and managers to handle functions, such as recruiting, hiring, employee onboarding, payroll management, employee reviews, paid time off, and employee documentation. Having the information in an online system rather than in a file cabinet somewhere allows remote employees to obtain

---

29 Pamela Hinds and Brian Elliot, "WFH Doesn't Have to Dilute Your Corporate Culture," Harvard Business Review, February 2021, accessed April 3, 2023, https://hbr.org/2021/02/wfh-doesnt-have-to-dilute-your-corporate-culture.

information and communicate with managers and the human resource team easier.

There are a few things to consider when selecting an online human resource system.

- Does it handle all of the current HR functions?
- Is it easy for employees to communicate with the HR team and managers?
- Does it provide self-service for employees?
- Will it make HR processes more efficient?
- Is it secure, and does it keep the data confidential?
- Does it integrate with any of your other systems?

A good virtual HR system will allow your employees to get the information they need easily, allow your HR personnel to communicate effectively with the team, and help keep your organization compliant with applicable regulations and laws.

Micro-Bonus Systems

In an office environment, it is easy for supervisors to recognize employees publicly or privately for good work. Coworkers can thank each other for assistance on tasks and projects. The interaction is natural and human. People want to be recognized and build good relationships with those who they work with. "Studies show that social connections play a central role in fostering a sense of purpose and well-being in the workplace. They also impact the bottom line: Effective management of social capital within organizations facilitates learning and knowledge sharing, increases employee retention and engagement, reduces burnout, sparks innovation, and improves employee and organizational performance."[30]

Micro-bonus systems are a company-sponsored recognition system that allows anyone to bonus another employee for work that they have done. The recognition has a small monetary value. More importantly, all the employees can see the recognition and notice the work that others are doing. It helps bond teams and fosters a desire for all team members to do better work.

Micro-bonuses can have negative consequences. Dr Monica Franco-Santos argues that the bonuses could be detrimental to organizations that have underlying issues. "These include difficulty preserving fairness, the potential for increased conflict between colleagues due to perceived biases and unfairness, and the encouragement of materialistic values. This kind of system can also generate gaming behaviors, encourage interest-based alliances between colleagues, and create additional stress and anxiety in the workplace."[31]

I believe that in the right circumstances, micro-bonus systems can enhance a workplace, especially if most of the employees are remote and cannot *see* the work that others are doing. It is a way for coworkers to bring great work to light and managers to publicly recognize the great performance of team members.

Group Games and Time to Unwind

Can't get together anymore after work to just hang out? Water cooler talk doesn't happen virtually. While some leaders may be excited by the increased productivity of the elimination of office small talk,

30 Rob Cross, *To Be Happier at Work, Invest More in Your Relationships,* July 30, 2019, https://hbr.org/2019/07/to-be-happier-at-work-invest-more-in-your-relationships (accessed July 25, 2022).

31 Cranfield University, "Peer to peer bonuses may have unintended negative consequences, expert warns," Phys.org, May 2019, accessed April 3, 2023, https://phys.org/news/2019-05-peer-to-peer-bonuses-unintended-negative-consequences.html.

Simon Sinek, the author of *Start with Why*, argues that "white space" is essential for team members to build strong relationships.[32] With the elimination of the office, creating a place the team members to interact outside of mandatory meetings, work projects, emails, and instant messages is essential.

> **With the elimination of the office, creating a place the team members to interact outside of mandatory meetings, work projects, emails, and instant messages is essential.**

How many times have you joined a video conference, and everyone continues to work until the meeting starts? Small talk isn't natural during online meetings like it is for in-person meetings. A leader needs to create opportunities to build strong relationships. As Simon Sinek explains, strong relationships aren't built in grandiose events but in small interactions between team members. Although more difficult virtually, there are ways that I have found that can build similar bonds.

One simple way I have found, as mentioned previously, is to start meetings with a segue where each person shares some good news. This can lead to additional conversations about what is going on in each other's lives.

There are also some great online interactive games, such as Jackbox Games. As a team, we've played these games over Zoom or Teams. It gives us time to laugh, chat, and learn more about each other's personalities.

Another idea is to use a platform such as MoveSpring to create company health challenges. Teams can compete against each other for

32 Simon Sinek, *Start with Why*, (New York: Portfolio, 2011).

the most steps over a period of time. It helps build team camaraderie and healthy habits, especially for those who work from home.

A final simple suggestion is to periodically leave a video session open as a group while everyone works. It gives an opportunity to chat, ask questions, and a sense that you are in the same location.

Whatever you do, using technology to bring your team together will help build cohesion inside your organization. It takes intention from the leaders to make it happen.

Workforce Management Checkup

Give yourself a score of 1 to 3 in each area. Add them up to see how you did overall in this area. Focus on just one area to improve upon for now.

AREA OF CONCERN	SCORE
My company is prepared to allow employees, where possible, to work wherever they are.	
My company has sufficient cybersecurity in place to protect the company's data and assets in a virtual environment.	
My employees can effectively communicate with each other.	
Employees can access necessary business data from any location, and that data is secure.	
I actively engage my virtual team members to make sure they are part of an intentional company culture.	
TOTAL **TOTAL POSSIBLE 15**	
NOTES	

Workforce Management Key Points

- Most companies have moved some or most employees to remote-only or hybrid work options.
- There are two types of communication: asynchronous and synchronous. Teams that successfully work remotely know when and how to use each type of communication.
- Modern cloud solutions allow teams to collaborate on the files and documents simultaneously.
- For companies that have non-cloud-based, line-of-business applications, solutions such as WVD allow them to still allow employees to access the systems wherever they work.
- Companies can implement technologies to provide better team engagement for remote and hybrid workers.
- Small talk and white space can still happen in a virtual environment; leaders must intentionally give opportunities for employees to build strong relationships.

CHAPTER 10

BUSINESS ACCOUNTING SYSTEMS

Accounting is the score card of business performance. Setting up your business accounting systems correctly is essential to the successful long-term performance of your business. Doing it incorrectly can cause your business grave damage. Even profitable businesses can run out of cash. Moreover, lack of good accounting systems can blind you to what is happening in the business, allow bad employees to cause you harm, or even good employees to make mistakes that can cost you a fortune. I have seen it all over my years in business consulting.

Let me give you an example of how bad accounting systems and processes can cause significant damage to your company. A few years ago, Verne Harnish, an author of several business books and a prolific speaker, was giving a business presentation in Russia. According to his article detailing the cyber-attack, he was practicing bank balance accounting—receiving daily alerts from his bank on his balances. He also had a trusted assistant whom he would communicate with over email, and she had the authority to perform wire transfers on her own without secondary approval. The morning that he was speaking to three thousand CEOs in Moscow, his email was hacked. He believes that it happened because he was connected on an unsecured wireless network. The hackers monitored his email correspondence, which included the large bank balances. By email, he had also instructed his assistant to wire some funds the day before. The hackers also saw that email and repeated a similar email to his assistant, instructing her to wire out $400,000 in three separate transactions. Because they had access to his email, they confirmed with her through his email that was what he wanted her to do. The bank flagged one of the transactions and attempted to contact Verne, but of course, the assistant intercepted that email and acknowledged the transfer because the hackers (posing as Verne) told her to do so over email. Because Verne had a very lax accounting policy, hackers were able to take $400,000 from his company.[33]

Take the time to set up the right accounting systems in your business, and the dividends will pay off in the future.

This story could easily be in the cybersecurity section of this book, but I am including it here as a reminder of the need to *set up your accounting systems and processes correctly.* Even small businesses have potentially large sums of money that can be accessed by hackers. You may not have $400,000 sitting in your bank account, but what if because of poor accounting policies, a disgruntled employee was

33 Verne Harnish, *$400k Defrauded; Cyberattacked; Expensive Lessons; Don't Let it Happen to You!*, October 13, 2016, https://verneharnish.typepad.com/growthguy/2016/10/400k-defrauded-cyberattacked-expensive-lessons-dont-let-it-happen-to-you.html (accessed August 18, 2022).

taking $5,000 from you every month for several years? Those dollars could add up over time to $400,000. I've also seen that happen to small businesses. That is money that you will most likely never see again. Take the time to set up the right accounting systems in your business, and the dividends will pay off in the future.

Another aspect to good accounting practices is being able to see into the future. As we discussed in chapter 2, creating budgets and pro formas is essential to creating a vision for the future of your enterprise. Those budgets and pro formas will be more accurate if the historical information that they are based on is correct. Banks, investors, and eventually buyers like to see that a company is forecasting accurately and hitting or exceeding their forecasted revenues and profits. Companies that can see into the future accurately are worth more than one that dreams of riches but never achieves them.

I'll tell you a slightly funny story about planning, or lack thereof, and the impact it can have. In long-distance hiking, the backpacker must carry all his own food, which requires planning and decision-making of weight, space, and calorie requirements. Freeze-dried food is the easiest to acquire and use for long trips, but the taste leaves much to be desired. We joke on the trail that all freeze-dried food generally tastes the same; it starts with the same distinct freeze-dried base flavor, and a few ingredients are added to make it taste a little less gross. There are alternatives to freeze-dried food, but they require much more planning. Because I am generally rushing from one thing to the next, I tend to just buy the freeze-dried food and put it in my pack. After a few days on the trail, I dream of real food from home or even anything non-freeze-dried. I start eating energy bars for meals rather than the planned freeze-dried meals.

At the end of each hike, I tell myself I need to plan better next time. Do I actually do that? Not usually. A few hours of meal planning would save me much pain on the trail, but I often choose the convenience and pain over a little extra time in preparation and having the chance to enjoy better meals on the trail. However, lack of planning for the trail cuisine doesn't have too negative an outcome on my hike, just my taste buds! But for your business, don't be lax with your accounting, as the result will have much more impact; instead, be sure to plan, prepare for everything you can in advance, and enjoy your business outcomes!

In this chapter we'll briefly review some accounting system options available for small and midsize businesses and discuss how to choose a system for your company and how to manage the accounting systems and processes in your company.

Key Definitions

- **Business intelligence (BI):** Software that provides dashboard, analysis, and reporting information from multiple data sources, such as finance, operations, sales, and marketing.
- **Customer relationship management (CRM):** A software system that contains information on your customers and contacts. It allows you to have continual and effective communication with them, including marketing, sales, and service delivery functions.
- **General ledger:** The list of accounts in your business, their transactions, and balances. It is the core of your accounting system.
- **Material requirements planning (MRP):** A system that allows manufacturers or other products-based businesses to know what raw materials and components are required to build, package, and deliver the final products to the customers.
- **Enterprise resource planning (ERP):** A complete management system that includes general ledger accounting, customer relationship management, manufacturing resource planning, operations management, and service delivery functions.
- **Lean:** A system, generally used in manufacturing, to eliminate waste, increase productivity, and improve delivery time of products and services.
- **Profit and loss statement:** Also known as an income statement. It shows your sales, costs of goods sold, gross profit, general expenses, and net profit for a period of time (month, quarter, year).

- **Balance sheet statement:** A listing of general ledger accounts, grouped by assets, liabilities, and capital, with their associated balances on a specific date.
- **Cash flow statement:** A listing of the cash and cash equivalents entering and leaving the business, generally divided into operations, investing, and financing activities. It gives a clear picture of the cash at the start of the period, what happened to it during the period, and what is left at the end.
- **Generally accepted accounting principles (GAAP):** A common set of accounting rules, guidelines, and principles issued by a governing body. Following GAAP allows companies to have consistent and comparable financial statements.
- **Personally identifiable information (PII):** Information that permits an individual to be identified by either direct or indirect means.

Choosing a system

Your accounting system is the core of your financial and operation systems. Choosing the right system requires you to have a vision for your company. Many systems work well for small businesses but do not scale to larger enterprises. If you already have a system in place, converting to a new platform generally requires a major project and will lead to loss of historical information. There are some specific considerations you should take into account when choosing a new system.

1. Cloud-based vs. internally hosted: If your business operates out of a single location and you do not plan on growing, you can easily store your data on one computer or a server. If you ever plan on having multiple locations, having the accounting system stored in one location may cause issues in the future. In that case going with a cloud-based accounting system may be an advantage.

2. Basic accounting vs. ERP: For most small businesses, having a system that provides a basic general ledger and financial reporting is sufficient. As your organization grows, having integration with other business functions, such as sales, marketing, manufacturing, and operations may become *more* important. An ERP system will allow all those functions to integrate together in one system.

3. Single vs. multiple entities: Small business accounting systems do not handle multiple entities very well. If you have one location or one operation, a basic accounting system will work. If you are planning on expanding or opening multiple operating entities, an ERP system may make more sense.

Although not a comprehensive list, these are some common accounting systems that you may consider for your company:

Type	Name	Best Use
Cloud	Intuit QuickBooks Online	Small businesses and nonprofits with less than fifty employees. Good for hybrid/remote companies.
Server/ Desktop	Intuit QuickBooks Desktop	Small businesses that require more features or integrations than Quick-Books online. Requires data to be stored on a server or desktop.
Cloud	Xero	Micro-businesses that just need basic accounting software.
Cloud	FreshBooks	Good for service-based small businesses. It has good integrations with many apps.
Cloud	Sage Accounting	Formerly Peachtree. For small businesses that are looking for an alternative to QuickBooks.
Cloud	Sage Intacct	Advanced financial software for medium-sized businesses or multiple entities.
Cloud	Microsoft Dynamics 365	Complete ERP system for financial, CRM, sales, marketing, and HR.
Cloud	Oracle NetSuite	Advanced ERP system for financial, CRM, sales, marketing, and HR.

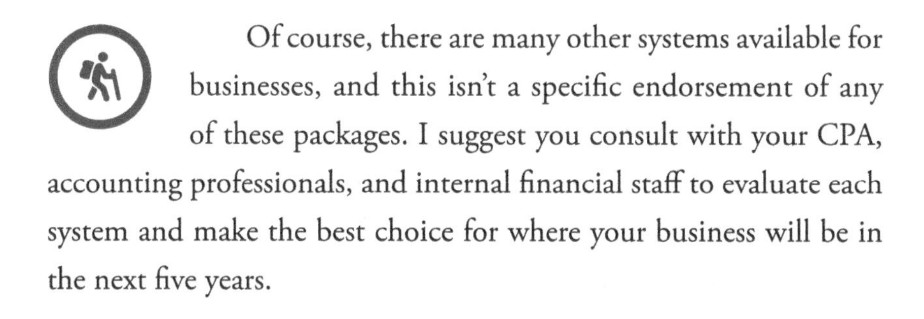

Of course, there are many other systems available for businesses, and this isn't a specific endorsement of any of these packages. I suggest you consult with your CPA, accounting professionals, and internal financial staff to evaluate each system and make the best choice for where your business will be in the next five years.

Reporting

Since accounting functions as a rearview mirror for your business, it should produce a scorecard. There are three reports that, as a business owner, you need to be able to read, understand, and act upon. Having these reports completed and reviewed monthly will give you a picture of the status of your business. Additionally, comparing the current results against historical data and your future pro forma will let you know if your business is on track or offtrack.

1. Profit and loss statement: This is also commonly referred to as an income statement. The report starts with your sales. Your costs of goods sold, or your variable costs to produce the sales, are subtracted from the sales to produce your gross profit. Finally, all other expenses, generally fixed costs such as facilities and administration, are subtracted to produce your net income. It is said that the top line of this report is for vanity, and the bottom line is for sanity. Both numbers should be positive and growing. When you pull the reports, look at three different periods: monthly, quarterly, and annually. This will give you trends and perspectives. A common error I've seen is pulling reports for partial or different periods. This type of reporting will give you skewed results.

2. Balance sheet: This report is a snapshot in time. It should be reported on at the end of a specific period: month, quarter, or year. It shows your company's assets, liabilities, capital, accumulated income, and book value. I've found that most business owners have a more difficult time understanding the balance sheet than an income statement. Have your financial advisor or CPA review the balance sheet with you on a regular basis and look at trends over time. It will give you an opportunity to make decisions on where to invest in your company and the results of those investments.
3. Cash flow statement: Even profitable businesses can run out of cash. How can this happen? There are expenditures in a business that do not show up on the profit and loss statement. The principal portion of any debt payments, capital purchases that are depreciated, increase in inventory or accounts receivable, and distributions are all non-profit-loss cash drains on the business. Closely monitoring your cash flow statement will help you understand the results of all your actions in the business.

Dashboards

Bob Parsons, founder of GoDaddy.com, stated in his "16 Rules for Success" that "anything that is measured and watched, improves."[34] One of the principles that I like from the book *Traction* is the idea

34 Bob Parsons, *My 16 Rules For Success*. n.d., https://bobparsons.com/16-rules/ (accessed August 26, 2022).

that "everyone has a number"[35] that is measured and tracked. It gives each team member a clear idea of their goals.

Using a dashboard system will allow you to view everyone's number and other specific measurables that the business operates from. These measurables could include operational, financial, sales, and marketing data. I like the use of dashboards because they give management a quick look at current status and trends. They can make decisions based on those trends and make quick adjustments rather than having to wait for monthly, quarterly, or annual reports to be produced.

Often referred to as business intelligence software, these systems allow you to connect to multiple data sources, combining information into one place. We call this type of solution bringing your data into one "pane of glass" because you can review the most important information in your business from a single system. Current software options for dashboards include the following:

- Domo
- Tableau
- BrightGauge
- Sisense
- Dundas
- Microsoft Power BI
- Zoho Analytics
- Yellowfin

35 Gino Wickman, *Traction: Get a Grip on Your Business,* (Dallas, Texas: BenBella Books, Inc, 2011).

Choosing a dashboard depends on the size of your team, the data sources that it will be connecting to, and the complexity of the data. As with most software, I'd recommended reviewing several options, getting demos of the software, and working with an implementation team to set the software up correctly in your environment. You'll also need someone or a group of people on your team who maintains the software. Dashboards need maintenance to keep the information up to date with any changes to operations, systems, or business functions.

Monitoring

As a business owner or executive officer, it is your responsibility to monitor the finances of the company closely. Cash is the lifeblood of the business, so any depletion of cash can cause a crisis. As we discussed in chapter 6, even well-intentioned employees can make mistakes. I have been called in to assist companies too many times that have had money wired out to hackers, invoices paid to wrong vendors, or employees cheating the company a few dollars at a time.

Taking time on a periodic basis (I'd suggest monthly) to review in detail the financial position of your organization will help prevent fraud. If employees know you are monitoring the money closely, there is less likelihood of fraud.

You can use technology to assist in this. For example, one online payment system allows controls over payments to vendors. Each bill requires one or more approvals before payments. You can attach the actual invoice into the system so each person who approves payments can look at the invoice and verify the amounts match. This also prevents vendors from being paid that the company may not use anymore. Each department can verify payments and make sure that the services being paid for are still valid. How many times have you paid a vendor for years only to realize that you are not using their product or service anymore? It's a terrible feeling to know you've thrown money away. Use technology to assist you in saving money!

Two-Factor Accounting

Speaking of monitoring your accounts, wire and ACH fraud are some of the most prevalent ways hackers use to get money from companies. Just like the Vern Harnish story, they will get an executive to click on a link and gain access to their emails. They will monitor how they approve payments and wait for a long time before making a move. They'll use the same formats to request a wire to be made and send an email off to a well-meaning employee to make a transfer. Everything will appear normal to that person, and they'll make the transfer, giving away money to illegitimate recipients.

I suggest that no payments, wires, or ACHs leave your company without a second type of verification. Most banks will now provide a two-factor key for ACH and wires. Only a limited number of people in the organization should have the ability to be a secondary approver of payments. The person making the payment from the request of an executive should verify the request by a second form of communication. Email should never be solely relied on for sending money. There is too much fraud in email. Verbal authentication should confirm the payments, along with a nonwritten passcode. This second step will significantly prevent fraud in your organization.

The reasons for internal fraud are generally associated with what is called the fraud triangle: opportunity, pressure, and rationalization. Most theft is caused when all three exist. An organization generally cannot control the pressure or rationalization on an employee but can reduce the opportunity by making fraud as difficult as possible. By making the act of fraud very difficult and the chances of being caught as high as possible, a company can prevent most financial fraud.

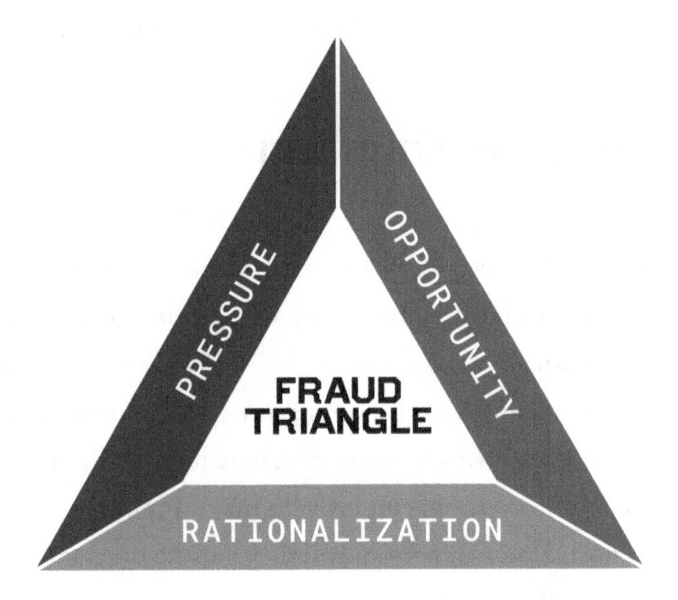

Here are a few ways you can use two-factor accounting to reduce the risks of fraud:

- The owner reviews the company's bank statements monthly.
- Match vendor payments with vendor invoices.
- Verify all account changes, such as addresses or ACH payment accounts, with vendors through means other than how the change request came.
- Split the positions of accounts payable, accounts receivable, and account reconciliations.
- Limit the number of signers on all accounts.

- Know that authorizations for ACH and wire transfers only occur with multifactor authentication to the bank account and by a separate person who creates the payments.
- Implement positive pay on all bank accounts.

There are many other ways to make fraud as difficult as possible. I would suggest you discuss this topic with your CPA and security professionals and implement a plan to put their suggestions in place.

Personally Identifiable Information

All companies hold a significant amount of PII, and it is the responsibility of every person inside the organization to protect that data. PII is highly regulated, and we will cover some of those regulations in chapter 11. Examples of PII contained inside your company include the following:

- Employee information: Social security number, date of birth, passport information, home address, medical information, payroll data, and bank account information
- Customer data: Credit cards, bank account information, federal and state tax IDs, addresses
- Biometric data: Fingerprint scans, retina scans, voice signatures, and geometric facial data.
- Human resource data: Drug screening, background checks, driving records, employment records, and emergency contact information
- Patient data (for medical practices): Age, gender, ethnicity, health history, medicines, allergies, immunization status,

laboratory test results, discharge instructions, and billing information

Safeguarding this data comes down to individual responsibility to ensure that the data is protected, is only used for needed purposes, and is not exposed to individuals or organizations that do not need it or are not allowed to have access to it. Ultimately, there must be one person in the company who can attest that this data is protected.

In chapter 3 we discussed the need for employee training. One of the biggest areas of concern is protecting PII inside the organization. A large part of employee training should be centered around protecting this information, which includes identification, classification, securing, and destroying PII. Regulations require that any breach of PII must be reported to the regulating authority and individuals affected. Your company should review its PII policies often to ensure that the data is protected according to the regulations at the minimum or set a standard higher than regulating authority. Loss of this information can be extremely costly to the organization in fines, fees, loss of reputation, and loss of business. I recommend using an outside company or organization to periodically review your company's handling of PII to ensure that you are doing everything required to protect it.

I recommend using an outside company or organization to periodically review your company's handling of PII to ensure that you are doing everything required to protect it.

Business Accounting Systems Checkup

Give yourself a score of 1 to 3 in each area. Add them up to see how you did overall in this area. Focus on just one area to improve upon for now.

AREA OF CONCERN	SCORE
My company uses an accounting system that best matches the company's size and growth trajectory.	
The leadership team receives and reviews regular financial reporting.	
Each employee has a key metric that they are responsible for, and all metrics are monitored and reported.	
The company has implemented two-factor accounting to ensure the company's money is protected.	
PII is identified, classified, and properly secured across all areas of the company.	
TOTAL **TOTAL POSSIBLE 15**	
NOTES	

Business Accounting Systems Key Points

- Different types of accounting systems exist. Choosing the right accounting system is important to provide proper accounting procedures and financial reporting.
- Advanced accounting systems such as ERP systems can allow for easier management of financial information in growing organizations.
- The company's leadership should regularly review three key reports to show business health: the profit and loss statement, balance sheet, and cash flow statement.
- Dashboards allow managers to quickly view data and keep track of key metrics inside the organization.
- Each employee should know and be able to monitor a key metric that they are responsible for.
- Two-factor accounting protects the company's finances and prevents fraud.
- Every company has PII about their employees and customers.
- Protecting the company's PII is ultimately each employee's responsibility.
- Companies need to invest in training employees on how to identify, classify, and secure PII.

CHAPTER 11

LAWS AND COMPLIANCE

Throughout the book, I've discussed different ways to keep your company safe in a highly complicated world. Technology has brought significant advancements in productivity, connectivity, and speed of doing business. Companies store and process information each day that can compromise individuals and other businesses. Because of this, government agencies have passed laws to protect others from the conduct of businesses. These laws are ever-changing, so this chapter is just an introduction to the types of laws that exist today, mostly in the United States. This chapter is different in format than the rest of the book. Each section will give a brief overview of a law or regulation, who the regulation affects, and where you can find more information about it.

Payment Card Industry Data Security Standard (PCI DSS)

What Is PCI DSS?

PCI DSS is a set of standards created by payment card processors that all credit card processors must follow in order to accept, process, or store credit card information. It is an industry standard and not a government regulation.

Who Is Affected?

Any company that accepts, processes, or stores credit card information. Even if you use a third party to process charges, your company is affected by these industry rules.

Where Can You Find More Information?

Visit the industry website: https://www.pcisecuritystandards.org/

What Happens If You Are Not Compliant?

Because PCI DSS is an industry standard and not government regulation, credit card processors are generally the ones that will impose fines, fees, or other consequences related to noncompliance. Usually, organizations that are not compliant pay higher processing fees and a monthly noncompliance fee and open themselves to civil law suits for potential release of credit card information.

How Do You Become Compliant?

Start by speaking with your credit card processor. There are different levels of compliance requirements depending on how much you process and if you store credit card information. Generally, your company goes through an annual certification and attestation. If you are out of compliance with the requirements, your processor will charge you higher fees. There are companies and consultants that can assist you with becoming compliant, thus reducing your costs of processing credit card payments.

National Institute of Standards and Technology Cybersecurity Framework

What Is the NIST Cybersecurity Framework?

National Institute of Standards and Technology (NIST) Cybersecurity Framework is a set of standards and recommendations designed to help organizations to reduce cybersecurity risk.

Who Is Affected?

All organizations can voluntarily follow the NIST framework. A large portion of this book is dedicated to many of the principles found in the framework. By executive order, all United States federal government agencies are required to follow the framework. If you are a government contractor, you may be required to follow the framework. [36]

Where Can You Find More Information?

Visit the NIS website: https://www.nist.gov/cyberframework

What Happens If You Are Not Compliant?

---

36 Executive Office of the President, *Strengthening the Cybersecurity of Federal Networks and Critical Infrastructure*, May 16, 2017, https://www.federalregister.gov/documents/2017/05/16/2017-10004/strengthening-the-cybersecurity-of-federal-networks-and-critical-infrastructure (accessed September 09, 2022).

If you are not a federal government agency, there are no requirements to implement the framework. Compliance with the framework will help your organization to have better protections against cybersecurity attacks.

How Do You Become Compliant?

Start by retaking the digital assessment in chapter 1 of this book. As you implement the recommendations in this book, you will be improving your cybersecurity footprint. Review the frequently asked questions on the NIST website. Additionally, a professional can perform a security assessment of your organization to discover your compliance with the framework and provide recommendations to implement.

HIPAA and HITECH

What Is HIPAA and HITECH?

The Health Insurance Portability and Accountability Act (HIPAA) and the Health Information Technology for Economic and Clinical Health (HITECH) Act are part of two United States federal laws that require medical organizations to protect patient healthcare records and information. Although not entirely about digital compliance, because most healthcare information is stored electronically, electronic compliance is a large part of HIPAA and HITECH.

Who is affected?

Any individual or organization that processes, stores, or handles healthcare records. Additionally, any organization that is a vendor must have a business associate agreement with the healthcare provider.

Special note: I have found that groups such as dentists or small medical practices think that they do not need to be compliant. This is not true. Any organization that holds protected health information (PHI) or other health records, no matter the size or the organization, is required to implement and follow the rules of HIPAA and HITECH.

Where Can You Find More Information?

Visit the HHS website: https://www.hhs.gov/hipaa/for-professionals/index.html

What Happens If You Are Not Compliant?

In the event of a breach or release of PHI, the fines can be extensive. Health and Human Services (HHS)[37] is authorized to fine up to $1.5 million, which does not include state fines. California, for example, has imposed a $25,000 per patient record that is "unlawfully or without authorization accessed, used, or disclosed."[38]

How Do You Become Compliant?

The best course of action to become compliant is to hire a HIPAA compliance expert to assist in the compliance process. The experts will perform a risk analysis, conduct routing monitoring of access and record keeping practices, implement access controls, assist with technical implementations of electronic safeguards, and train employees on how to protect PHI. Even in my company, we use

37 Health and Human Services, *HITECH Act Enforcement Interim Final Rule,* June 16, 2017, https://www.hhs.gov/hipaa/for-professionals/special-topics/hitech-act-enforcement-interim-final-rule/index.html (accessed Septelber 12, 2022).

38 California Legilative Information, *HEALTH AND SAFETY CODE - HSC,* January 1, 2015, https://leginfo.legislature.ca.gov/faces/codes_displaySection.xhtml?lawCode=HSC§ionNum=1280.15 (accessed September 12, 2022).

an outside agency to assist our clients with compliance. It requires a level of expertise and familiarity with the laws and regulations to help organizations become fully compliant.

FTC Safeguards Rule for Tax Preparers

What Is the FTC Safeguards Rule?

The Federal Trade Commission (FTC) has the authority to create rules to safeguard taxpayer data. The Internal Revenue Service (IRS) has published two documents to assist tax professionals in complying with the safeguard rule. The intent is to protect tax and financial information of individuals and corporations.

Who Is Affected?

Any tax preparer or organization that prepares tax returns.

Where Can You Find Out More Information?

The IRS has created the following two publications that assist tax preparers to become compliant with the FTC safeguards rule:

Publication 4557: https://www.irs.gov/pub/irs-pdf/p4557.pdf
Publication 1345: https://www.irs.gov/pub/irs-pdf/p1345.pdf

What Happens If You Are Not Compliant?

As with HIPAA, the federal government can bring civil penalties against an individual or organization that violates the safeguards rule. The penalties for each violation or release of tax information in 2022 is $46,517.[39]

39 Federal Trade Commission, *FTC Publishes Inflation-Adjusted Civil Penalty Amounts for 2022*, January 6, 2022, https://www.ftc.gov/news-events/news/press-releases/2022/01/ftc-publishes-inflation-adjusted-civil-penalty-amounts-2022 (accessed September 12, 2022).

How Do You Become Compliant?

You can follow the recommendations of the two publications by the IRS to become compliant. This includes having a written security plan, having strong internal controls, encrypting all data stored on the computers, encrypting data that is transmitted and received from clients, and providing employee training. Does any of this sound familiar? Additionally, any suspected data theft or loss of client data should be reported immediately to the IRS. As with most things, attempting to cover up your problems is a worse violation than being compliant and then admitting to theft or loss.

CMMC

What Is CMMC?

Cybersecurity Maturity Model Certification (CMMC) is a framework developed by the United States Department of Defense to protect the government and its contractors (also known as the Defense Industrial Base) from cyberattacks. It is specifically designed to protect sensitive government and defense information critical to national security.

Who Is Affected?

Companies that receive federal contracts from the United States Department of Defense are required to comply with the CMMC. There are tiers of compliance, which minimally require an annual self-assessment up to a government-led assessment and attestation. The level of compliance depends on the size of the organization and contractor and the level set forth in the contract received by the organization.

Where Can You Find Out More Information?

Visit the Department of Defense (DoD) website on CMMC: https://dodcio.defense.gov/CMMC/

What Happens If You Are Not Compliant?

The DoD has specified that all contractors must be compliant by 2025. Eligibility to bid and receive DoD contracts depends on the company's compliance with CMMC.

Additionally, compliance with the model will make your company more secure and less vulnerable to cyberattacks. The benefits of compliance far outweigh any cost of noncompliance.

How Do You Become Compliant?

CMMC is a complex set of requirements, and I strongly recommend that you receive outside assistance to become compliant. Additionally, if you are required to be compliant with level two or three, you will need third-party assessments. To start, the DoD website has resources and materials to assist you in the compliance process. If you are a DoD contractor, you should be working on compliance before bidding on any contract.

Red Flags Rule

What Is the Red Flags Rule?

The Red Flags Rule requires businesses and organizations to implement identity theft programs to protect consumer information (PII), detect any warning signs, and report identity theft to anyone affected by it.

Who Is Affected?

Although the Red Flags Rule applies mostly to financial institutions, the application of the rule is broad enough that many businesses may be required to comply. If your company receives credit information or extends credit to its customers, you must comply with the Red Flags Rule.

Where Can You Find Out More Information?

Visit the Federal Trade Commission's website, which provides information on the rule: https://www.ftc.gov/business-guidance/privacy-security/red-flags-rule

What Happens If You Are Not Compliant?

The fines for noncompliance have a maximum penalty of $3,500 per violation or $2,500 per infraction. Because the violation is per record, the fines can get very large. For example, Vivint Smart Homes, Inc. settled with the FTC for $20 million for violation of the Red Flags Rule.[40]

How Do You Become Compliant?

The Red Flags Rule requires that a covered company must have a written identity security plan. The FTC website defines compliance in four steps.

- Identify relevant Red Flags.
- Detect Red Flags.

40 Federal Trade Commission, *Vivint Smart Home, Inc.* April 29, 2021, https://www.ftc.gov/legal library/browse/cases-proceedings/192-3060-vivint-smart-home-inc (accessed September 19, 2022).

- Prevent and mitigate identity theft.
- Update the program.

With most of these rules, I recommend you work with an expert to assist you in writing, implementing, and maintaining your program.

Laws and Compliance Key Points

- Implementing a written security plan is a requirement for most government regulations.
- Most businesses are covered by some type of organizational or government rule for protecting PII, PHI, or other confidential information.
- Good cybersecurity policies, covered in this book, are the basis for compliance.
- Laws are updated and changed constantly.
- Compliance often requires bringing in experts to assist you with developing your programs.
- Having third-party assessments is often required to ensure compliance with the written security program.

CHAPTER 12

DOCUMENTATION, POLICIES, AND PROCEDURES

In this trail guide, I've discussed many important topics that should change the way that you operate your business. I've personally read many business books to help me change the way I run my own business. Many of those books have literally gone in one ear and out the other. (Or is that in one eye and out the other?) It isn't until I have taken something that I've learned from a book and created a new policy or procedure inside my business that the information I learned made any difference. More importantly, I had to document the new way of doing things and train, train, retrain my team on it before it was a permanent change. As we close out this journey together, we'll discuss how to create policies and procedures around technology, or anything you want to do in your business, and make them stick.

As people, we are creatures of habit. We tend to get up in the morning at the same time, have similar routines to get ready in the morning, drive similar routes to work, and perform our tasks in the same way. To change the way we do things takes much more effort. In his book *The Power of Habit*, Charles Duhigg describes how we use habits and routines each day to save brain power. "Habits, scientists say, emerge because the brain is constantly looking for ways to save effort. Left to its own devices, the brain will try to make any routine into a habit, because habits allow our minds to ramp down more often. This effort-saving instinct is a huge advantage. An efficient brain … allows us to stop thinking constantly about basic behaviors, so we can devote mental energy to inventing."[41]

Changing any habit is a difficult task because it requires much more brain energy. Have you ever tried to change the way things are done in your business, just to get frustrated because everyone reverts to the way that it has always been done? I sure have! I go to a conference, read a book, have a brilliant idea, and then bring it back to my team only to be met with eye rolls and feet dug in. People don't want to change because it requires the brain to process so much new information, all while being tasked to perform at the same level that previously required less brain effort. Making changes in your business is hard because it *is* hard to change! It's hard to change because our brains are wired to find the easiest path to get things done, which is generally a previous routine or habit.

So how do you change policy or procedure (institutional habit) inside your company? First, understanding that habits occur because of the habit loop. First, a cue triggers the routine. Once the routine is performed, there is a reward. That reward could be anything, but

---

41 Charles Duhigg, *The Power of Habit*, (New York: Random House Trade Paperbacks, 2014).

it generally causes a chemical reaction in our brain that releases a bunch of dopamine. This is the brain's reward to the body for a job well done and makes us feel happy. To change a habit, first you will have to identify the cues that trigger the routine. Once the cue is identified, your brain (or your employee's brain) will have to work a little harder to recognize the cue and change the routine attached with that cue to the new one. After the new routine is performed, there must be some type of reward. If the old routine is performed, there should be some type of negative reaction to identify that the wrong routine was performed.

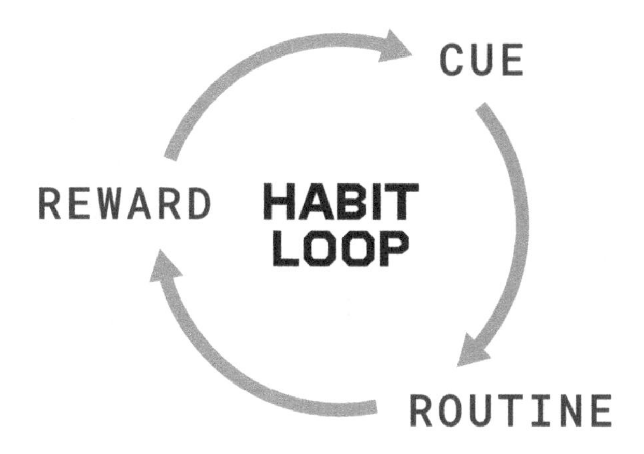

For a simple example in my personal life, when I started to exercise, it was difficult for me to get into the habit of doing something each day. I realized that because of my busy schedule during the day, I would need to start working out in the morning. I was not a morning person. (Okay, I still am not, but I fake it pretty well.) The start of my exercise habit was rough because just the act of getting up in the morning was punishment rather than a reward. Two things made a big difference for me: creating a social bond with friends and a reward of food. I started working out with a group of people I liked, including a good friend. We would meet early for a run, bike, or swim. Most of the bike and swim workouts were in a group. The reward was new friendships and big kudos from a coach who helped us improve. All this interaction and achievement releases dopamine and oxytocin, which was my first reward for getting up early and exercising.

Because I was doing so much aerobic exercise, my metabolism increased, and I started to burn calories much more easily. That allowed me to eat some extra food without weight gain (although I was losing weight at the time). I did my best to manage a healthy diet but would give myself a reward of a few Oreos or ice cream at night. I looked forward to that reward, if and only if I did the work in the morning. It was a little motivation and fed the inner craving for sugar. I now say I'm a fat kid with an exercise problem. Those two rewards allowed me to change the habits that were engrained in me for over thirty years. Now I still exercise (and eat Oreos at night), but I don't need the motivation as much. I've built the habit to wake up early every morning. When I don't get to work out, I feel off. A few good rewards created a lifelong habit that will be hard to break.

Think about that in your teams as you try to make change happen in your business. You'll need to do more than just dictate these policies because it requires a change in culture and habits. How are you going to reward your team (or yourself) for making positive changes in your organization? Are they motivated by money? Extra time off? A pat on the back? Oreos? Whatever it is, make it meaningful to them so they are willing to develop the habit.

Key Definitions

- **Documentation:** A written method of describing an application, process, purpose, information, or architecture. It should be sufficient that another person reviewing it can gain the same understanding of what was documented as the person who originally wrote it and provides useful information as to how to operate or use what was documented.
- **Document control system:** An electronic system that provides a method to store, track changes, process approvals, restrict access, and notify of required changes and updates.
- **Process:** A series of actions to accomplish a task and lead to a specific end.
- **Procedure:** A specific method to accomplish a task. A procedure should be clear and specific so that it can be trained and repeated with exactness.
- **Checklist:** A list of actions that must be accomplished, generally in order, to complete a specific procedure.
- **Process flow diagram:** A graphical sequential representation of a process or procedure, including its operations, timelines, options, and members.
- **Policy:** A course or specific action that is directed by the company to lead to a specific goal.
- **Muscle memory:** A procedure that has been repeated with sufficient frequency so that it can be accomplished with little conscious thought.
- **Routine:** A procedure that has been repeated with sufficient frequency as to allow it to be accomplished with expertise.

Writing Documentation

In my business, lack of documentation is death. For example, if a technician installs a new router, sets a complicated password, and doesn't document it, when another technician goes to work on that router later, they will not be able to log in to it. The only recourse would be to reset the router and start over from scratch, completely wasting the original technician's time and effort and potentially upsetting the customer.

Do you have similar instances in your business? If a person leaves your company, can someone pick up exactly where they left off and do the same work, or is there significant loss of knowledge that is hard to replace?

Documentation of systems, processes, procedures, and all other aspects of your business adds significant value to your company. It reduces the pain of turnover, provides structure, and eventually allows someone else to step into your company (hopefully after paying you a large sum for the business) and run it just as successfully as you did. Another big benefit of documenting your company is that it allows you to refine the systems, process, and procedures and produce greater efficiencies. Finally, it provides a way to train all staff to do the work the same way every time so that onboarding new team members is easy, and you don't have to hire "geniuses" to work for you. An average employee can accomplish difficult tasks because they have a documented recipe to follow.

Documentation of systems, processes, procedures, and all other aspects of your business adds significant value to your company.

So how do you document your business? First, I suggest some prework. There should be different classifications of documentation. You'll need to protect certain documentation from the outside world for legal, compliance, or proprietary reasons. Other documentation can be shared easily with all your team or publicly. Here are some different classifications that I would suggest for your documentation. When you create different documents, label them so they can be stored and protected appropriately.

CLASSIFICATION	EXAMPLES
Publicly accessible	General business information, marketing materials, and sales collateral
Confidential	Business processes, procedures, employee handbooks, and financial reports
Confidential with full encryption	Personally identifiable information, patient health records, and credit card information
Restricted access with full encryption	Passwords, logins, employee data, and accounting data
Highly restricted with full encryption	Intellectual property, proprietary information, and trade secrets

Back to my example of a technician documenting a client's router password. That would be classified as restricted access with full encryption. Not all employees of my company need access to that information, and we need to store that encrypted so that it cannot be retrieved by a hacker. Additionally, our storage system will track who accessed the password. In the event of that employee leaving the organization, we can reset it so that our former technician cannot access client data. I'll discuss writing these types of procedures in a subsequent section.

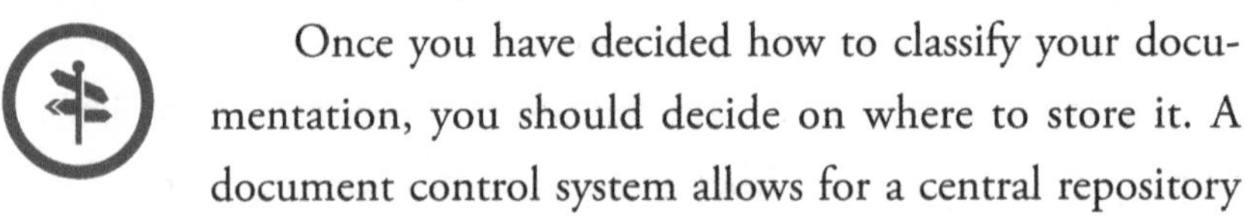

Once you have decided how to classify your documentation, you should decide on where to store it. A document control system allows for a central repository to store it, control access, track any changes, and have a system for approvals. It also provides a system for compliance, such as HIPAA or ISO 9001. Most document management systems provide the ability to create workflows that will notify of any changes, required updates, and approvals. Three examples of document management systems are M-Files, Maxo, and Microsoft SharePoint. Designing and maintaining a document management system is beyond the scope of this book. As I've often said, consult with a technology expert to review the systems and see what is best for your company. I don't recommend storing documentation on a shared drive because it is very difficult to track changes and restrict access on that system of storage.

Along with classifying your documents, I recommend creating a numbering or cataloging system. For example, divide your documentation into major sections and number them. Then for each major category, define subcategories. For example,

1. Business Operations
    a. Company vision, values, and mission
    b. Business units
    c. Company workflows
2. Technical Operations
    a. Widget manufacturing
    b. Servicing widgets

3. Finance
    a. How to count beans
    b. Accounts payable
    c. Accounts receivable
4. Sales
    a. Pricing
    b. Sales process
5. Marketing
    a. Marketing plan
    b. Website
    c. Social media
6. Human Resources
    a. Employee onboarding
    b. Payroll
    c. Expense reporting
7. IT Operations
    a. Security
    b. Systems

Your categorization and subcategorization are going to be much more extensive. Creating a list of everything you want to document will provide a basis for everything you need to document inside your business. The idea is to create a manual for how your business is run so that anyone can easily be trained to do the same process each time and with complete accuracy. This process can be overwhelming to start but will lead to a far more efficient and profitable (and sellable) business.

Documentation should be a living part of your company, which is why I have recommended a control system. You should build in timely reviews of your documentation to make sure the processes are valid. How many times have you written a how-to manual to just put it on the shelf and it gather dust? When you finally pull it out of the cobweb-filled corner, everything in it has changed. Take the time to develop a process to review, update, and validate the documentation.

For more information and a documentation template, visit https://nathanwhittacre.com/resources.

Password Management

Handling passwords is one of the most difficult parts of technology for people. They need to be complex, never repeated, highly protected, and ever changing. There is no way we can keep track of the hundreds of different passwords we've used in our brains. Compound that with institutional passwords—passwords and access that may need to be shared with others—and all the security that comes along with using passwords goes out the window. I've seen passwords shared on sticky notes on monitors, Excel files, Word documents, etc., all of which are generally unencrypted and accessible by too many people. Institutional passwords also tend to be repeated between programs and websites.

I strongly suggest the following two things for password management:

- Limit the sharing of passwords as much as possible. If systems allow you to create user accounts for individual people, do so. For example, a surveillance system may have one administrative password that was created when the system was set up. Most allow you to create users in the system rather than giving out the administrative user to everyone. It is much easier to delete a user when they leave the company than resetting the shared password each time an employee leaves the company.

- Use a password management system that allows for encrypted storage of passwords. Systems such as LastPass or 1Password allow for your employees to manage their passwords securely and share them if needed. They also can store those multifactor authentication tokens (the one-time passwords that come up in an app) rather than having them stored on a cellular device that can be lost or stolen.

Leaving password management up to your employees to figure out is very dangerous. I've found that they are going to go to the least painful option, which often involves sticky notes and Word documents. Your company should have a policy for password storage, which includes mandatory multifactor authentication. Ideally, you provide the system that all your employees use so that it is easy to train and verify compliance.

When our company completes audits and assessments of computer systems, we scan for password storage on the devices. We also look for passwords stored on the device that can easily be decrypted. All passwords stored in many web browsers, such as Chrome, Firefox, Safari, or Edge, can be decrypted by a program that does not need administrative privileges. That means that any software that is running on the company, legitimate or nefarious, can access any passwords that your employees save in the browser. I suggest turning off the ability for your employees to save passwords in the browser because it leaves a significant hole in your company's security.

Intellectual Property

Most people think of patents or copyrights when I mention intellectual property (IP). There is so much more to IP than patents. Your documentation, processes, and procedures are all intellectual property. Talk to any investment banker and they will tell you that the more that you have your IP well documented and protected, the more valuable your business becomes. You need to safeguard your IP by restricting access to confidential information, ensure that it cannot be stolen by outsiders or employees as they leave the company, and protect it by contract with your team members.

To spark some ideas about what you should protect, here is a list of items I would recommend that you document and protect inside your company:

- All processes and workflows
- Customer lists and demographics
- Former customers and prospect lists
- Marketing content and workflows
- Service or manufacturing processes (your business's recipes)
- Vendor lists and contacts
- Pricing and sales processes
- Lending and financial information
- Line of business software
- Web domains and content
- Social media access and content

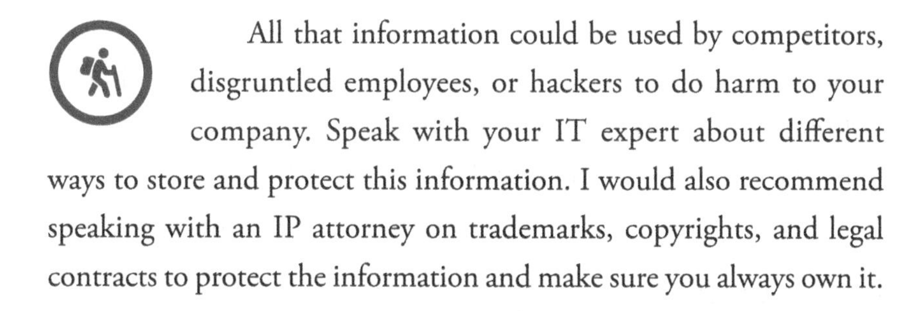

All that information could be used by competitors, disgruntled employees, or hackers to do harm to your company. Speak with your IT expert about different ways to store and protect this information. I would also recommend speaking with an IP attorney on trademarks, copyrights, and legal contracts to protect the information and make sure you always own it.

Processes and Procedures

I'm sure that you are not the only company in the world that does what you do. What makes you different from your competition is how you do it. As an entrepreneur, one of the difficulties I have had in my own business is making sure that each person inside my organization would do the work the same way I would do it. Getting to the point that everyone on your team producing the work, communicating with the customers, and even acting the same way is difficult but not impossible. Documenting processes and procedures is the starting point to achieving that goal.

Early in my business, I would communicate these processes and procedures verbally to my team. We would talk through how I would like the work to be accomplished. This process, in turn, would require my team to ask questions repeatedly, which would frustrate me. We'd also tell each other stories about how we did the work in the past. One day one of my team members said it was like we were a tribe that would pass down traditions from one generation to another through song. When a new "tribe" member would join the team, we would have to teach them the song so they could learn the "Stimulus Way." In a small organization, this may be effective but certainly is not scalable. We had to start documenting our processes and procedures so that anyone could follow them. We created a book and appropriately named it the *Tribal Songbook*. The large three-ring binder contained detailed instructions on how we did the work, including technical work, human resource management, accounting, and sales. It was the turning point in our ability to scale the business. We've long since updated the book and moved it to a documentation management system. As our business changes, we consistently update the documentation to make it relevant and accurate to how we want the processes to be completed.

Giving your team a visual representation of how a process works helps them understand and follow it.

I recommend using a tool like Microsoft Visio or Lucid Charts to diagram the process workflow. Giving your team a visual representation of how a process works helps them understand and follow it. You can also create decision trees to have different scenarios in the process. For example, the process flow attached is part of my company's ticket triage process and a decision tree about client acceptance of billing time on a ticket. You'll notice that in each step, the employee entering the ticket can make a decision based on the information they receive from the client, and it has a specific process for completing the ticket entry after a correct decision is made. In this example, if a ticket is being entered, it is what we call a break/fix agreement with a client (something breaks and we fix it for a fee out of contract) that has an estimated four hours to complete the work, and the customer service representative must receive written client approval before proceeding. The process is very clear, and the customer service representative doesn't need to think outside the box on each ticket entry. So long as each scenario is thought of in the process creation, it can be trained, followed, and executed with ease.

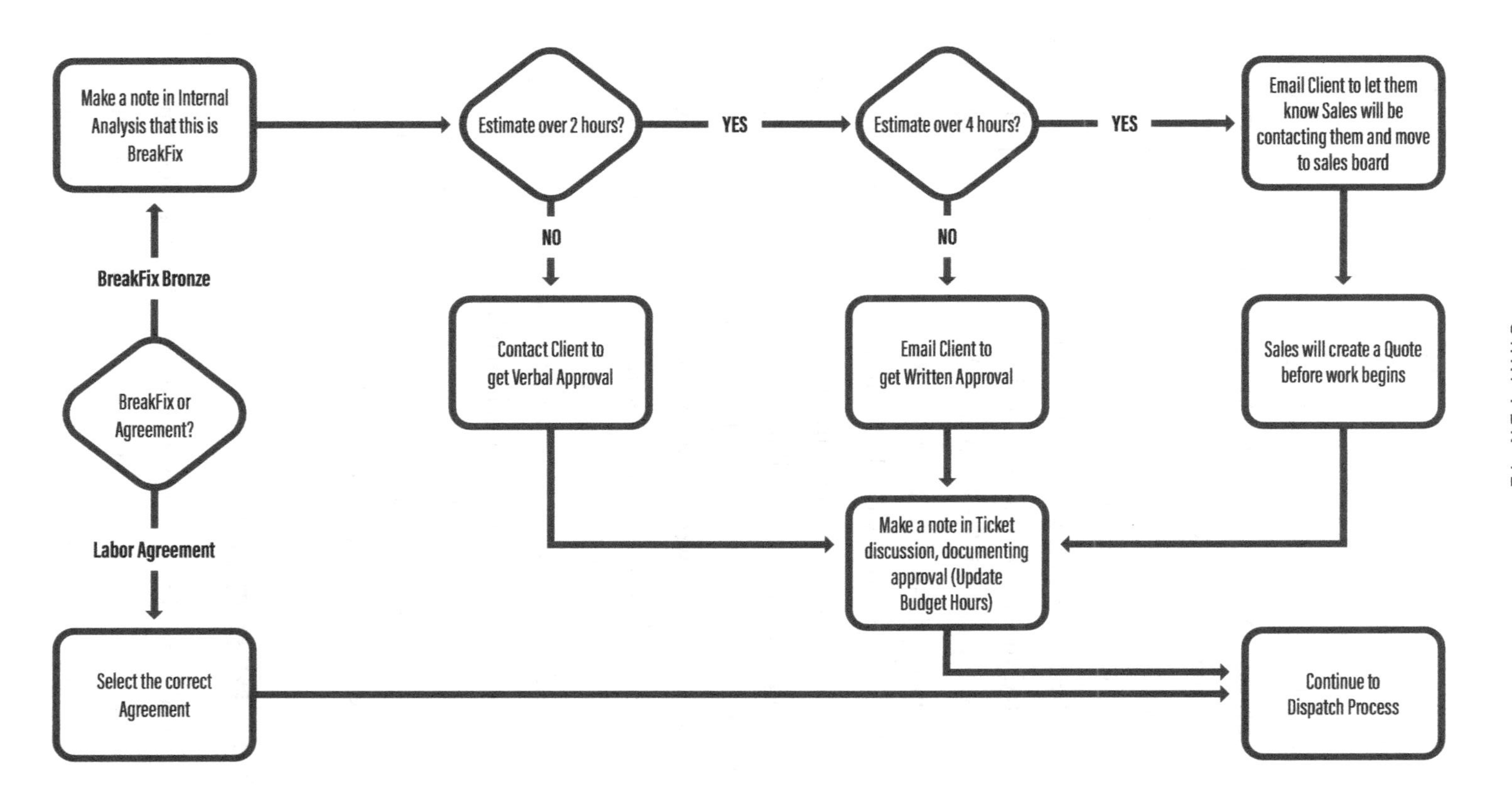
Make a note in Internal Analysis that this is BreakFix
Estimate over 2 hours?
YES
Estimate over 4 hours?
YES
Email Client to let them know Sales will be contacting them and move to sales board
BreakFix Bronze
NO
NO
BreakFix or Agreement?
Contact Client to get Verbal Approval
Email Client to get Written Approval
Sales will create a Quote before work begins
Labor Agreement
Make a note in Ticket discussion, documenting approval (Update Budget Hours)
Select the correct Agreement
Continue to Dispatch Process

Along with the process flow diagram, I recommend writing a long description of the process, with each step detailed in long form. That document should give the background as to why the process should be followed, how to follow it, and what the intended outcome of the process should be. For this process workflow, we have written the following example to illustrate *why* we need to follow the process:

If you think back to the last time you took your car to a repair shop, they went to great measures to provide you a written estimate prior to beginning any work. Just think how mad you would be if they quoted you $100 for a tune-up, but when you go to pick up your car, they give you a bill for $500, and they say, "Well, it needed a new water pump and fan belt, so we went ahead and installed them for you." I'm guessing this would not fly with you. That's right, if they found it needed those things, they would stop working and contact you to find out if you were willing to authorize the extra repairs. We need to conduct ourselves with the same respect for our clients.

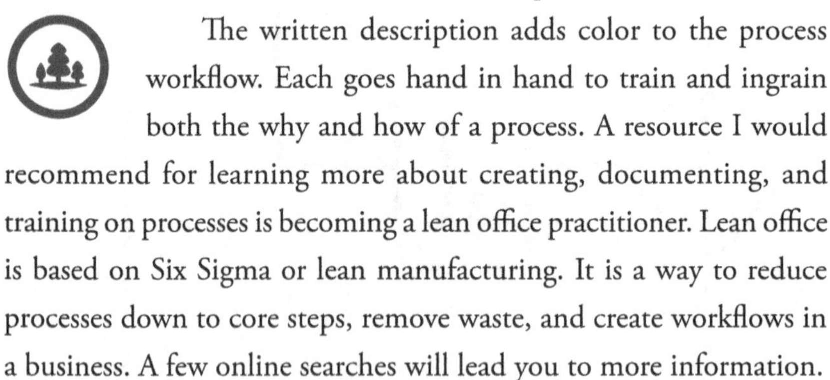

The written description adds color to the process workflow. Each goes hand in hand to train and ingrain both the why and how of a process. A resource I would recommend for learning more about creating, documenting, and training on processes is becoming a lean office practitioner. Lean office is based on Six Sigma or lean manufacturing. It is a way to reduce processes down to core steps, remove waste, and create workflows in a business. A few online searches will lead you to more information.

The Champion

On a final note about processes and procedures, I recommend you have a champion inside your organization for each process. This person should be the expert around the process. Not just the expert but also the true believer or evangelist of the process. To ensure that a process lives well past the date of creation, someone inside the company needs to promote, train, update, and understand the *why*. When the process is challenged, which it always will be, the champion will overcome all obstacles and guarantee that it is followed by the organization.

Documentation, Policies, and Procedures Checkup

Give yourself a score of 1 to 3 in each area. Add them up to see how you did overall in this area. Focus on just one area to improve upon for now.

AREA OF CONCERN	SCORE
My company stores core processes and procedures in a document management system.	
Documentation is classified by levels of access.	
My company's intellectual property is protected from misuse and theft.	
A password manager is used by all employees, and password policies are enforced across the entire organization.	
Each core process inside my organization has a champion that ensures compliance.	
TOTAL **TOTAL POSSIBLE 15**	
NOTES	

Documentation, Policies, and Procedures Key Points

- Writing documentation can be a difficult task but ensures that company knowledge exists outside the heads of the employees.
- Documentation of core processes and procedures increases the value of the business.
- An enterprise password management system can reduce risk to the organization of stolen credentials.
- Storing passwords in a web browser is dangerous because it can be accessed by any software running on the computer.
- All businesses have intellectual property that requires protection.
- Using process flow diagrams makes following a process easier to train and understand.
- Give the *why* behind each process to improve understanding and compliance.
- A champion can ensure compliance to processes across the whole organization.

AFTERWORD

We've been through a long journey to get to here. My hope is that you have enjoyed the trip so far. As I have written this book, I have realized how much information that I've presented and how overwhelming it can feel. In my research I have learned things myself and changed the way that I think about business, technology, and security. My hope is that you have learned along the way too.

You may be overwhelmed by all this. I've heard many times from business owners that they want to focus on what they do well and not technology. I get that. You are an expert in your profession, and that is why you have customers and clients that look to you to solve their problems. Going back to the beginning of the book, my one hope is that this book has enlightened you on a few subjects so that you can make better decisions in your business and communicate with technology experts about how to incorporate these ideas into your company.

So what do you do now with all this power and knowledge? My suggestion is to go back to one chapter of the book that impacted you the most and retake the checkup. Try to improve one or two questions over a three-month period. Make some goals and give assignments to your management team, and have them report back to you weekly on the progress. At the end of the quarter, your company will be one step better.

I've found that setting three-month goals is better than yearlong goals. It is enough time to get something meaningful done but not too long that you lose sight of the objective. After four quarters, you will look back at the year and see significant progress in your business.

If there are any critical areas in which you need to make fast and significant progress, I would suggest engaging with outside experts who can assist you. Especially in areas of cybersecurity, IT infrastructure, and networking, working with these experts can guarantee proper implementation and faster results than attempting to handle it internally. Even for me, I use outside experts to ensure that my company is implementing technology correctly and that we are protected from the latest threats.

Finally, make sure you are documenting your new processes and procedures. Use the documentation to train and retrain your team on the new way of running your business. Creating a library (your own tribal songbook) of how you run your business is more valuable than you can imagine!

Don't give up on your digital journey. One step at a time will lead you to your destination. Digital success will improve your business, take you beyond your competition, and lead you to long-term business and financial security. I'm here for you to guide you along. You can always reach out to me for expert advice and trail guidance. Visit http://www.nathanwhittacre.com to message me and to find additional resources. My hope is that this book inspired you along your business and digital journey.

Happy trails!

EPILOGUE

It was the summer of 2017, and I was standing on the trail in Yosemite National Park with a group of Boy Scouts enjoying the beautiful scenery. I saw a sign that read "John Muir Trail - 211 miles." I had never seen or heard of the John Muir Trail (JMT), but I was intrigued. I had been an active Boy Scout growing up and attended many scouting activities, campouts, and other outdoor adventures, eventually obtaining my Eagle Scout rank when I was seventeen. During those years in scouting, I had never gone on a long backpacking trip before. The following year I was turning forty years old and wanted to do something very different to celebrate that time on this earth. For the last seven years, I turned from an overweight computer nerd to a triathlete, marathon runner, and cyclist, but that wasn't enough, in my mind. I needed something more adventurous for my fortieth birthday. A total of 211 miles of backcountry trekking was much more intriguing.

After the hike with the boys and when I had cellular service again, I looked up information on the trail. It was 211 miles starting at Happy Isles Trailhead in Yosemite Valley all the way to the peak of Mount Whitney, which is 14,505 feet above sea level, the highest point in the continental United States. Over the expanse of 211 miles, the trail goes over eight mountain passes and almost 40,000 total feet of climbing. It also passes mountain lakes, snowcapped peaks, and some of the most beautiful country in the United States. Running through the High Sierra mountains, the trail is one of the most coveted hikes in the United States. This, I thought, was what I wanted to do to celebrate my fortieth birthday.

When I got home, I looked up what it takes to get a permit to hike the trail. The demand for the JMT is so high that the National Park Service had to create a lottery system to provide permits to hike the JMT. On any given day during the lottery, only 1–2 percent of applicants would receive their permit to hike. My chances weren't great to get my permit to hike, but I decided to try anyway. I began researching which time of year was best to hike the trail. To me, August or September 2018 would be the best time for the hike. Snow mostly would have melted on the trail. There are not as many problems with bugs or difficult river crossings because much of the snow would have already melted. It's one of the least traveled times, especially through the middle of September.

After reading many blog articles, websites, and several books on the trail, I started the process of getting a permit through the lottery system. I also started researching what it takes to hike the wilderness: estimate the number of miles a day possible, determine where to camp, resupply, and so forth. Based on averages, I estimated that I would take somewhere between eighteen and twenty-one days to complete the hike. Luckily, I was in very good shape, so I assumed that I could make even better time on the days I felt well. I was going to have to take off work and be completely out of cellular and Internet range, cutting off communications with the outside world for the entire time. This started to sound like the adventure that I wanted to take for my fortieth birthday!

Since this was my first thru-hike, I decided to find some experts to get information on what it takes to live in the wilderness for several weeks. I decided to pay a visit to one of my favorite places. REI is like a grown-up version of Toys "R" Us. I could spend hours in that store looking at all the different things and dreaming of the adventures I could have using them. From biking to running, hiking to climbing, camping to kayaking, the store was my idea of a playground. When I visited the local store to see what information they had about the JMT and planning for a long thru-hike, amazingly, the person whom I talked to had hiked the JMT several times in the past. One time he did it with a group and another time completed it solo. Spending well over an hour discussing the trail with the REI representative gave me significant insight on how to be able to do the hike on my own, with a group, or finding a group to do it with. That information was invaluable for planning and deciding who I was going to hike with.

I ran around the store, taking notes and pictures of the equipment that he recommended. Even more importantly, I made mental notes of things that he said he tried and that didn't work out well for him. I started to create a list of things that I would need to buy. The list was significant because I didn't have anything for backpacking. Most of my gear was for camping and rather large and bulky. In my mind I imagined trying to haul a propane stove, a six-man tent, and a Dutch oven for 211 miles. That wasn't going to work. I appreciated the long list of equipment and supplies because it was expert advice from somebody who had done it in the past several times. Sometimes knowing what not to do or buy is just as valuable as what to do.

I started thinking that I needed advice from more people about the trail. Talking to people who had done the hike in the past and hearing about their experiences offered a gold mine of information. One researcher compiles a list each year from surveys that people send him about the trail. Things such as equipment weight, miles traveled per day, problems that forced them to leave the trail, and general trail conditions analyzed by time of year were very important to planning. I also joined a social media group that allowed people to ask questions about the trail. I get nervous about asking questions online because you can get so much disinformation and opinions that conflict with each other. I found that in this group, there were several moderators who were very experienced backpackers, including two who had written books about the trail. Being able to ask some questions when I got stuck on planning proved invaluable. Many of the responses weren't informative, but someone usually had some additional insight that I hadn't thought about.

Part of my planning included reading the book *Wild* by Cheryl Strayed. She hiked part of the Pacific Crest Trail (PCT) as part of her change of life. If you've read her book or watched the movie, you'll remember that she also took a bunch of advice from the local REI and ended up buying everything, literally everything, so much that she couldn't fit it all in her pack. More importantly, she couldn't even lift her pack on her back. She eventually met some people on the trial who helped her reduce her supplies to something more manageable. She found that it was about balance, figuring out what was absolutely necessary for the hike and what was just a want. That is part of thru-hiking: figuring out the necessities and then choosing a few lightweight wants that will make the trail more comfortable. I found a book on the PCT written by Ray Jardine, where the author described reducing weight to an absolute minimum. He advised sawing off the end of a toothbrush to reduce its weight. I knew that I didn't want to go to those extremes, but I took some of the advice into consideration, especially the realization that everything I brought would have to be carried for all 211 miles!

Food is one of the biggest challenges on the trail. There are no stores or any place to stop to grab some extra stuff. In the first half of the trail, there are a few "pack stations" that have some supplies and where food can be shipped. On the second half of the trail, there is nothing near the trail. To get off the trail, it is often at least a one-day hike in and out to get to any of the closest cities. You often must hitchhike from the trailhead into the city. It might take you two days to get to the city and then back on the trail. The black bear calls the Sierra its home, and they tend to prefer hikers' candy bars rather than the berries they find on the trees. For this reason, hikers must carry all their food in bear canisters, which can only hold about ten days' worth of food. At some point during the hike, I would have to resupply my food provisions. There are several options to resupply along the JMT, and they are strategically placed at different points along the way, except for the second half of the trail, where the only option is to leave the trail to go to the nearest town. That meant that I would need to complete the second half in about ten days, which was doable in my mind.

One of the most popular resupply points is called Muir Trail Ranch (MTR), which is exactly halfway between Yosemite Valley and Mount Whitney. It is a few miles hike off the trail but much less than most of the other options. They have a good system of the hikers mailing a five-gallon bucket full of food to a PO Box, where they retrieve it with mules and haul it into the backcountry. They hold them at the ranch for crazy backpackers to pick up their food. A few weeks before I set out on my journey, I mailed my estimated amount of food for ten days on the trail to MTR and hoped it would all be waiting for me in a few weeks.

To provide permits for the JMT, the National Park Service has a lottery system that hikers enter each day. It consists of sending in an application that is good for three weeks on the lottery. Since only about 2 percent of lottery applicants get their permits on any given day, if the hiker doesn't get their permit on a particular day, they are automatically entered into the next day's lottery. Each day the park service sends the hiker an email, letting them know that if their application was accepted or denied. The lottery happens about six months before the anticipated start date of the hike. I started submitting my lottery application in February, which was about six months before my planned start date. Each day I would get a response in my email box at about nine o'clock in the morning, denying me the permit. It was a terrible way to start the day for over three weeks. I started to get really disheartened because if I started late in the season, let's say the middle of September, the weather would get bad, especially heading into October. I knew that if I did not get a permit soon, I would not be able to go on the hike, and all the planning and preparation would have been for nothing.

One day I was sitting in a meeting with my business coach, and I received the usual morning email from the park service. It had a different subject line, which announced that I had received my permit! I was going to head out on the JMT to celebrate my fortieth birthday. With my accepted application in hand and much of the planning in my mind, it was time to start buying everything I needed. I also planned several test hikes to try out all the gear that I had purchased. I can't believe how many little things are needed to be able to do these hikes: water purifiers, cooking stoves, a lightweight sleeping bag, rain gear, first aid, hiking boots, and socks. Everything needed to be purchased specifically for hiking. Why? It had to be lightweight. It had to be comfortable for my body. If I wanted to do this hike quickly, carrying more than fifty pounds, including food and water, having it on my back was not going to work.

When I finished my shopping spree, I thought it would be prudent to try out the gear on a few shorter hikes before stepping onto a 211-mile trail in the middle of the mountains. I decided to do a three-day hike near Big Bear, California. The trail I chose intersected with and went along the PCT. The PCT is an approximately 2,600-mile trail that runs from Mexico to Canada along the mountain range near the western coast of the United States. It was built around the same time as its famous cousin, the Appalachian Trail (AT), which runs along the eastern coast of the United States. The PCT and JMT run along the same route for much of the JMT with a few exceptions. My hope was that the terrain, hiking, and camping would be similar in Big Bear along the PCT as along the JMT, and this proved to be the case. During the test hike, I realized that there were a few things that I still needed to buy to make the hike work. I also learned about some food that I liked on the trail and some that I didn't like so that I could purchase more of the stuff I liked and less of the stuff that I didn't. Over the summer of 2018, I did several more test hikes to make sure that all the equipment that I was using worked correctly and that I was able to hike the number of miles that I wanted each day. Those few test hikes were invaluable to making my JMT hike successful.

After all the planning and preparation, completing the hike became the easy part of the journey. I enjoyed seeing the amazing country that God gave us. It was so beautiful, so incredible, and one of the best experiences that I've ever had in my entire life. I grew closer to myself and closer to God and understood many things about myself that I hadn't had in the past. I enjoyed the difficult days of hiking, which added up to about 18 to 20 miles per day. Each day of walking on this backcountry trail, looking at amazing scenery in the mountains, hearing running water, and seeing the expanse of sky above made me feel very small in this world and at the same time very grateful to be part of it. Without the preparation beforehand, if I had just decided one day to throw on the backpack, with a few candy bars and a loaf of bread, and head out onto the trail, my experience would have been terrible. With loads of preparation, advice, and knowledge, I was able to do it solo without any issues.

Why am I telling you this story about my JMT hike in 2018 as an epilogue to a book on cybersecurity and computer technology?

The reason is simple: you have to be prepared to run a business.

Technology today is very complex and has many moving pieces. It is in every aspect of your business. Whether you are a sole proprietor or run a business with thousands of employees, I can assure you that without technology, you could not run your company. Technology is the lifeblood of your business; it runs in all parts and sustains it in a very competitive world. Business owners and management teams that ignore or delegate away the responsibility of technology will not survive.

Experts, knowledge, and the opportunity to use technology surround you. I decided to write this book as a handbook, much like those I used when planning my trek on the JMT. Use it as a reference guide, and you will be able to successfully navigate the digital age and the cyber world that we live in today.